Optogalvanic Spectroscopy

Sponsors

A G Electro-Optics Ltd
Chell Instruments Ltd
Coherent (UK) Ltd
Edinburgh Instruments Ltd
John Smith & Son (Glasgow) Ltd
Newport Ltd
Optometrics (UK) Ltd
Spectra-Physics Ltd
Spectroscopy Group of the IOP
The Scottish Development Agency
The United States Office of Naval Research European Office

Meeting Organiser

Dr R S Stewart

Optogalvanic Spectroscopy

Proceedings of the Second International Meeting on
Optogalvanic Spectroscopy and allied topics
held at Strathclyde University, Glasgow,
2–3 August 1990

Edited by R S Stewart and J E Lawler

Institute of Physics Conference Series Number 113
Institute of Physics, Bristol, Philadelphia and New York

Copyright © 1991 by IOP Publishing Ltd and individual contributors. All rights reserved. Multiple copying of the contents or parts thereof without permission is in breach of copyright but permission is hereby given to copy titles and abstracts of papers and names of authors. Permission is usually given upon written application to IOP Publishing Ltd to copy illustrations and short extracts from the text of individual contributions, provided that the source (and, where appropriate, the copyright) is acknowledged. Multiple copying is only permitted under the terms of the agreement between the Committee of Vice-Chancellors and Principals and the Copyright Licensing Agency. Authorisation to photocopy items for internal use, or the internal and personal use of specific clients in the USA, is granted by IOP Publishing Ltd for libraries and other users registered with the Copyright Clearance Center (CCC) Transactional Reporting Service, provided that the base fee of $3.50 per copy per article is paid direct to CCC, 27 Congress Street, Salem, MA 01970, USA.
0305-2346/91 $3.50 + .00

CODEN IPHSAC 113 1-338 (1991)

British Library Cataloguing in Publication Data

International Meeting on Optogalvanic Spectroscopy
(2nd 1990 Glasgow)
 Optogalvanic spectroscopy. — (Institute of
 Physics conference series)
 I. Title. II. Stewart, R. S. III. Lawler, J. E.
 535.8

ISBN 0-75498-047-4

Library of Congress Cataloging-in-Publication Data are available

This work relates to Department of the Navy Grant N00014-90-J-9026 issued by the Office of Naval Research European Office. The United States has a royalty-free license throughout the world in all copyrightable material contained herein.

Published under The Institute of Physics imprint by IOP Publishing Ltd
Techno House, Redcliffe Way, Bristol BS1 6NX, England
335 East 45th Street, New York, NY 10017-3483, USA
US Editorial Office: 1411 Walnut Street, Philadelphia, PA 19102, USA

Printed in Great Britain by Galliard (Printers) Ltd, Great Yarmouth, Norfolk

Preface

This two-day meeting was organised on behalf of the Spectroscopy Group of the Institute of Physics and held in the Department of Physics and Applied Physics, Strathclyde University, Glasgow, 2–3 August 1990. What better time to hold it than during Glasgow's year as the European City of Culture. The meeting demonstrated how understanding of Optogalvanic Spectroscopy has progressed since the previous meeting on this subject in Aussois, France (1983). The aim was to bring together researchers with interests over a range of complementary spectroscopic techniques and this proved to be highly successful. There were 45 participants of whom 15 were from overseas. The mix of youth and experience and wide range of expertise ensured much interaction and lively discussion.

There were 9 invited lectures (45 mins) and 19 contributed papers (25 mins). The review lectures were of a very high standard and the transcripts should prove an excellent basis for young scientists starting research in any of these areas. (John Goldsmith's lecture on multi-photon combustion diagnostics will be published elsewhere.) The contributed papers described much new material and many interesting techniques and applications which should be of interest to the wider scientific community. We are grateful to all the invited speakers and contributors.

We would also like to thank the following for their help in the organisation of the meeting: June McPhail and Jean Lindores for their very special efforts, June for much help with the meeting correspondence and Jean for her perseverence with the notices, programme, abstracts and editing, Ronal Brown, Cathy Calder and Tom Haggerty for help with the exhibition, Ken McKnight, Khalel Hamad, Colin Whitehead and Gordon Robb for excellent support during the meeting, Josephine and Gordon for turning the Stewart home into a meeting office and helping in many different ways, and most importantly, Edward Steers (Chairman of the Spectroscopy Group) for his ever-wise counsel.

It is a pleasure to acknowledge the financial assistance of all the sponsors whose support was vital to the success of the meeting.

R S Stewart
J E Lawler

Contents

v   Preface

Section 1: Molecules

1–26   Optogalvanic spectroscopy of molecules and complexes*
H H Telle†

Section 2: Plasma Sheaths

27–64   Laser studies of discharge plasma sheaths*
J E Lawler† and E A Den Hartog

65–70   Electric field measurement in the cathode sheath of a DC glow discharge in hydrogen
C Barbeau† and J Jolly

Section 3: Optogalvanic Modelling

71–80   Optogalvanic effects in very low current discharges*
R J M M Snijkers, G M W Kroesen, J M Freriks and F J de Hoog†

81–88   Optogalvanic detection of excited-state photoionization in the neon positive column
J Halewood† and R C Greenhow

89–107   Collisional mixing model for the optogalvanic effect on the 6402A ($1s_5-2p_9$) line in the neon positive column
R S Stewart†, K I Hamad and K W McKnight

Section 4: Resonance Radiation Trapping

109–132   Calculation of resonance radiation trapping*
F Vermeersch† and W Wieme

133–140   Experimental investigation of the imprisonment of the ($^3P_1-^1S_0$) 146.96 nm resonance radiation in Xenon
F Vermeersch, N Schoon†, E Desoppere and W Wieme

Section 5: Resonance Ionization Spectroscopy

141–158   Aspects of trace analysis using resonance ionization mass spectroscopy and some applications*
K W D Ledingham†

* Invited lecture
† Paper presenter

159–162 A spectroscopic study of aniline using resonance ionization mass spectrometry
  A Clark†, A Marshall, R Jennings, K W D Ledingham and R P Singhal

163–168 Resonance ionization mass spectrometry applied to the trace analysis of gold
  P T McCombes, I S Borthwick†, R Jennings, K W D Ledingham and R P Singhal

Section 6: High Resolution Spectroscopy and Novel Detection

169–190 New directions in high resolution optogalvanic spectroscopy*
  A Sasso†

191–196 Frequency entrainment in the relaxation oscillator method of optogalvanic spectroscopy
  J S Dunham†, S C Bennett and C O Butler

197–206 Isotope shift studies of Gd I transitions using laser optogalvanic spectroscopy
  T Pramila†

207–214 Optogalvanic spectroscopy in commercially available hollow cathode lamps
  M Duncan† and R Devonshire

Section 7: Plasma Diagnostics

215–231 Concentration profiling in a neon optogalvanic lamp using concentration-modulated absorption spectroscopy*
  T R Griffiths, W J Jones† and G Smith

233–240 Measurement of the copper vapour concentration in a pseudospark discharge by laser-induced fluorescence
  G Lins†

241–255 Laser-induced breakdown spectroscopy as an analytical tool
  B J Goddard, R M Allott† and H H Telle

257–262 Optogalvanic spectroscopy as a diagnostic of operating fluorescent lamp electrodes
  M Duncan† and R Devonshire

Section 8: Penning and Auto-ionization

263–282 Optogalvanic spectroscopy of plasma processes and autoionization levels*
  R Shuker† and M Hakham-Itzhaq

283–288 Two-step laser optogalvanic spectroscopy of strontium: Even parity, $J = 0$–3, autoionizing 4dnl Rydberg states
  A Jimoyiannis, A Bolovinos and P Tsekeris†

* Invited lecture
† Paper presenter

Section 9: Poster Papers

289–295 High resolution IMOGS measurements in heavy elements
D Ashkenasi†, G Klemz and H-D Kronfeldt

297–302 An original method for measuring the conductivity of low pressure discharges
R I Cherry† and T R Robinson

303–306 Ultra-trace analysis of NO by high resolution laser fluorescence and ionization spectroscopy
M Hippler†, A J Yates and J Pfab

307–316 Time-resolved fluorescence studies of the laser dye DCM
G Hungerford†, D J S Birch and R E Imhof

317–329 Radial model for the optogalvanic effect in the neon positive column
K W McKnight† and R S Stewart

331–336 Perturbations induced in CO–He–Xe discharges by laser oscillation
C E Little† and P G Browne

337 Author Index

† Paper presenter

Optogalvanic spectroscopy of molecules and complexes

H H Telle

Department of Physics, University College Swansea
Swansea SA2 8PP, United Kingdom

ABSTRACT: The technique of opto–galvanic spectroscopy is used for the investigation of molecules in flames and discharges. Besides the detection of stable molecules it has become possible to observe radicals, molecular ions and collisional complexes. Some examples, both for low resolution pulsed laser excitation and high–resolution cw excitation, will be described and advantages and limitations of the method will be discussed.

INTRODUCTION

Optogalvanic (OG) detection of atoms and molecules in plasma sources is being accomplished by perturbing their electrical equilibrium through interaction with radiation that is resonant with a transition of a species in the system under investigation. The OG detection is based on a change in the impedance or, equivalently, a variation of current in the plasma. The phenomenon was first observed for atoms by Foote and Mohler (1925) and Penning (1928), and for molecules by Terenin (1930). However, it needed the appearance of the laser to fully exploit the OG phenomenon for spectroscopy, and because of its simplicity, high sensitivity and selectivity the technique has successfully been used in a wide variety of laser spectroscopic experiments, including analytical flame spectroscopy (see e.g. Turk et al, 1979), laser stabilization (see e.g. Skolnick et al, 1970), plasma diagnostics (see e.g. Ausschnitt et al, 1978), atomic spectroscopy (see e.g. Goldsmith and Lawler, 1981), molecular spectroscopy (see e.g. Webster and Rettner, 1983) and dynamics (see e.g. Miron et al, 1979). Although the OG effect utilizing laser radiation was first observed for a molecule (a change of discharge current in a CO_2 laser tube was observed when the laser action was switched on or off; see Carswell and Wood, 1967) applications of the OG technique are mostly reported for atomic systems.

The mechanism of the OG effect has generally been thought to involve laser enhancement of the ionization rates of a particular species present in the plasma or a laser–induced depletion of the populations in metastable excited states. Several models have been developed, each appropriate for a given type of plasma, but they are principally phenomenological in nature and do not account for many of the complex processes occurring in ionized gases that could, in principle, be involved in a laser–induced impedance change. On the other hand, the generality of the OG effect is evident from the fact that a number of different plasma sources has been utilized, for example flames, positive column DC discharges, hollow cathode DC discharges, RF discharges and direct photo ionization.

Investigations performed on molecular discharges have shown that the OG effect can be observed for both electronic (see e.g. Feldmann, 1979) and vibrational (see e.g. Webster and Menzies, 1983) transitions. But other than in atoms in general it has been difficult to forecast the OG effect in molecules whose energy levels are very closely spaced and often do interact with each other. To provide a complete description of the OG effect observed for molecules also processes have to be included where the change in translational, rotational and vibrational temperature play a dominant role; however, often the knowledge of the discharge phenomena is not precise enough. The degradation of the laser excitation energy into translational temperature, sometimes also molecular dissociation, is evident when the resulting pressure change is monitored; this latter phenomenon is known as the optoacoustic effect. Thus ideally the optoacoustic effect, and furthermore fluorescence signals, should be recorded simultaneously to the OG spectrum to account for all the processes which may occur in the molecular plasma; this is shown in the form of a block diagram in Figure 1.

In a typical OG experiment a laser beam is introduced into a plasma which normally operates under equilibrium conditions. Upon laser excitation of a transition the state populations deviate from the equilibrium but after a certain time will asymptotically return to steady–state conditions. A non–equilibrium excited state population can perturb the plasma through a number of different processes. Qualitatively a change in plasma impedance will occur when these processes either change the charged particle density or mobility. Specific processes which must be considered are:

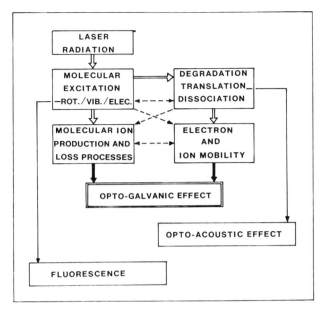

Figure 1: Block diagram of processes in discharges and flames resulting in the optogalvanic (OG) and optoacoustic (OA) effects.

(i) electron impact ionization: $X^* + e^- \to X^+ + 2e^-$

(ii) superelastic collisions: $X^* + e^-(\text{thermal}) \to X^+ + e^-(E \gg \text{thermal})$

(iii) associative ionization: $X^* + X \to X_2^+ + e^- + \Delta E$

(ΔE is usually small because the molecular ion is formed with highest probability in ro–vibrational states nearly resonant with the excitation energy)

(iv) Penning ionization: $X^* + Y^*(\text{metastable}) \to X^+ + Y + e^- + \Delta E$

(v) cathode photoemission: $X^* \to X + h\nu$

$$h\nu + M \to e^- + \Delta E$$

(if the energy of the photon is higher than the work function of the cathode material, an electron can be liberated).

The change in charged particle density, induced by the laser, is usually observed via the related voltage change, ΔV, across a ballast resistor or across the electrodes.

This (small) voltage signal is proportional to the population numbers, Δn_i, when the perturbation is small compared to the total population in the reservoir state; it then may be described by

$$\Delta V = -\beta \ \Sigma a_i \Delta n_i$$

with $\beta = [(\delta\alpha/\delta V)|_{n_i}]^{-1}$ and $a_i = (\delta\alpha/\delta n_i)|_V$; α is the number of generated electrons. The perturbation in the excited state population density caused by the laser will return to equilibrium with a time constant characteristic for the respective states. The dynamic behaviour of the system obeys the equation

$$d(\Delta n_i)/dt = \Sigma \gamma_{ij} \Delta n_j - \Sigma (n_i - n_j) \sigma_{ij} I_{ij}$$

where γ_{ij} are the rate coefficients for transitions between levels i and j, without laser radiation, σ_{ij} are the optical cross sections and I_{ij} are the light intensities. Only in a few (atomic) cases is it possible to reduce these sets of equations to simple two– or four–level models (see e.g. Erez et al, 1979). Furthermore, laser–induced changes in recombination rates, charge exchange or ion – molecule reactions are not being included in this simple model. For example, electron – ion recombination can produce the corresponding neutral species in a high Rydberg state which then relaxes and/or predissociates. While processes like these complicate the mathematical description significantly they can prove valuable for sensitive monitoring of phenomena which otherwise are difficult to trace, such as, for example, predissociation channels in highly excited molecules.

OPTOGALVANIC DETECTION OF MOLECULES IN FLAMES

In conventional spectroscopic techniques, used for analytical purposes, concentrations of molecules are being determined either by monitoring the absorption from the incident light beam or the re–emitted fluorescence. These techniques often suffer the inherent disadvantage of comparing two large signals, i.e. attempting to detect a small change in a large photon flux. In contrast, optogalvanic spectroscopy monitors signals against a much reduced background noise, determined mainly by the electrical stability of the flame (or discharge). Most studies in flames have concentrated on atomic species, and there are only relatively few publications on molecular species.

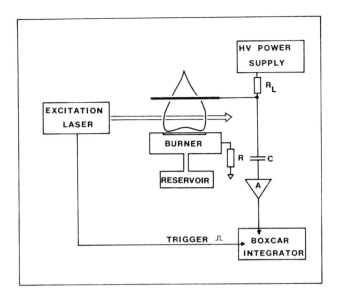

Figure 2: Schematic experimental arrangement for the detection of the OG effect in flames.

A typical apparatus to monitor the photon–induced current changes, as a function of laser wavelength, is shown in Figure 2. It consists of three main parts: (1) the burner – electrode assembly, (2) the tunable laser (pulsed or chopped cw) and (3) the data acquisition electronics. Depending on the molecular species to be investigated, for example stable gaseous molecules or combustion products, the flame is maintained at approximately atmospheric pressure using appropriately premixed gases. The burner itself forms one of the electrodes while the other electrode, usually 10–20 mm above the burner, consists of small rods placed symmetrically next to the zone in the flame which is illuminated by the laser. These latter electrodes are in general kept at a negative potential to extract the positive ions from the excitation region. The laser is usually adapted for the spectroscopic needs, i.e. when relatively low resolution is sufficient pulsed lasers are being used ($\Delta\nu \simeq$ 1–5 GHz) while high resolution requirements necessitate a narrow band cw laser ($\Delta\nu \simeq$ 0.001–1 GHz). In the former case the detection electronics incorporates gated boxcar integrator components while in the latter case the cw beam has to be chopped and a lock–in amplifier extracts the correct frequency component of the signal.

An example for the vast capabilities of molecular OG spectroscopy in flames is the investigation of metal oxides which are the result of combustion reactions within the flame. Schenck et al (1978) have recorded a number of transitions in rare earth oxides, and in particular lanthanum oxide (LaO) was investigated. In a typical experiment an aqueous solution of a metal compound, e.g. $La(NO_3)_3 \cdot 6H_2O$, at a concentration of 10^{-3}–10^{-2} molar is aspirated into the flame at rates of a few millilitre per minute. It is estimated that the premixed H_2/air flame has a temperature in the order of ~ 2000 K, and it is assumed that in its adiabatic environment the only products are water, nitrogen and excess hydrogen. This constitutes an environment in which electronic excitation of products, including LaO, is low compared to e.g. conventional arc sources, and thus it is reasonable to assume that laser excitation will mainly involve transitions which originate in the ground electronic state of the molecules. It should be noted that, even though electronic excitation is minimized, optogalvanic spectroscopy in flames is particularly amenable for the detection of high–lying electronic states; however, under the described experimental conditions no evidence of transitions between excited states was found for LaO.

The laser beam is usually directed along the axis of the flame, close to the top of the reaction zone, just a few millimeters below the negative electrode. Pulsed laser excitation (in Schenck's experiments laser pulses of 7 ns duration were used) is preferential over chopped cw excitation since in flames like that used here often large DC background currents are generated which may be completely saturated by the applied potential. The pulsed photon–induced current changes, produced by resonant absorption, are often very large and are easily processed by a boxcar integrator. The spectra exhibit excellent signal–to–noise ratios with a resolution limited by the laser linewidth (here 0.03 nm); a representative spectrum is displayed in Figure 3. The observed signals were linear with laser power indicating that ionization of the electronically excited molecules is due to collisions. The intensity depends on the lower state population and the collisional ionization probability of the excited state; collisional ionization becomes more efficient for lower energy separation between the upper state and the ionization continuum, and thus the OG sensitivity increases for transitions which reach highly excited electronic states. This is in agreement with findings for OG spectroscopy for atoms for which the strongest signals are usually observed when the population distribution in highly excited states is perturbed.

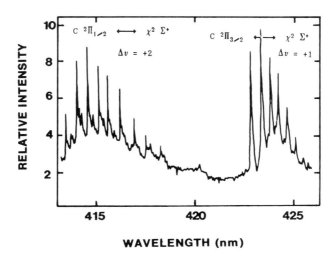

Figure 3: OG spectrum of the $C\ ^2\Pi - X\ ^2\Sigma^+$ transition of LaO, produced in a flame reaction; laser linewidth $\Delta\lambda \simeq 0.03$ nm (adapted from Schenck et al, 1978).

OPTOGALVANIC DETECTION OF MOLECULES IN DISCHARGES

Nearly any discharge type can be used in OG spectroscopy. For example, the first identifiable molecular structure was reported for bands of Cs_2 in a device in which the OG current flowed between a heated filament and an anode (Collins et al, 1973), and shortly after other alkali dimers were investigated in thermionic diodes (see e.g. Collins et al, 1976). However, most OG experiments have been performed in cold cathode discharges (DC glow discharges or hollow-cathode discharges) or in radio frequency (RF) discharges, either in inductive or capacitive coupling configurations. These principle discharge configurations are displayed in Figure 4.

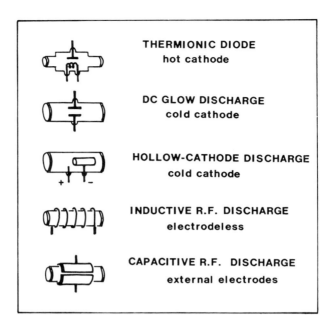

Figure 4: Schematic discharge configurations used in the detection of the OG effect.

Experimental arrangements for OG measurements vary widely, depending on the discharge device and the laser source. However, the general building blocks for the apparatus are common to all set-ups; thus only a typical example will be described in some detail. This is shown, schematically, in Figure 5 for a hollow-cathode discharge apparatus.

In this case the anode is connected through a ballast resistor, R, to a voltage-stabilized power supply, and the radiation-induced change to the discharge current is measured as the voltage drop across the electrodes. A capacitor, C, decouples the discharge from the detection electronics protecting it from the DC high voltage component of the discharge tube voltage. The light beam is mechanically chopped if a cw laser is used for excitation to provide the necessary intensity changes for the generation of the OG effect; temporal intensity changes are inherent to pulsed lasers. The OG signals then are either sampled via phase-sensitive detection (lock-in amplifiers) or gated detection (boxcar integrator), and appropriate sampling times are chosen to maximize the signal-to-noise ratio.

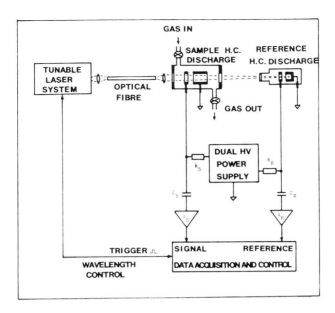

Figure 5: Block diagram of the experimental set–up for the detection of the OG effect in molecules, including wavelength calibration in a reference atomic species hollow–cathode lamp.

Typical operating conditions are a few hundred volts across the discharge tube, and the discharge current is of the order of 1–100 mA; the pressure inside the tube is a few mbar and has to be optimized for the molecular gas under investigation.

Laser powers of a few mW cw or a few μJ pulsed are sufficient to generate OG signals although it should be kept in mind that they may be rather weak since the size of the OG effect strongly depends on the population density of the particular states being perturbed by the laser radiation; with the inherently larger number of states in molecules the population in individual levels is much lower than in the atomic case. The spectral resolution is governed by the bandwidth of the laser when it is broader than the Doppler width of the molecular transitions; if the laser bandwidth is narrower than the molecular Doppler profile special techniques, like e.g. IMOGS, have to be used to reach sub–Doppler resolution.

The laser radiation, after passing through the discharge containing the molecular gas, is directed into a further hollow-cathode discharge, in the simplest case into a commercial spectral lamp; OG spectra of atomic species (rare gas buffer atoms exhibit the strongest signals) are used for wavelength calibration which is often vital for the assignment of features in the usually dense molecular spectra.

A number of molecules has been investigated over the past 15 years and the application of the OG effect in molecular spectroscopy will certainly grow because of its simplicity. In the sections below a few examples will be discussed which reveal the versatility of OG spectroscopy, including the analysis of molecular states at low and high resolution, the investigation of transient species and ions, and the possibility to extract parameters on dynamics in the discharge.

Investigation of stable gas molecules

A number of small molecules which are stable in the gas phase has been investigated; references may be found e.g. in Ernst and Inguscio (1988) or Daran (1989). In this paper the discussion will be restricted to the example of one diatomic molecule, namely the homonuclear molecule N_2. In Figure 6 partial spectra for N_2 are shown, obtained under different operating conditions for the discharge and using both pulsed and cw lasers; these spectra are compared to a normal emission spectrum at medium resolution.

The top of the figure shows part of the vibrational progression in the B $^3\Pi_u \rightarrow$ A $^3\Sigma_g^+$ transition in N_2. As can be seen the resolution of our monochromator ($\Delta\nu \sim$ 65 GHz) was insufficient to resolve the narrow rotational structure of the bands but the triplet nature in the sequence is clearly revealed. This rotational structure begins to show when a pulsed laser of medium bandwidth ($\Delta\nu \sim$ 3 GHz) irradiates the discharge volume and its wavelength is scanned appropriately (see e.g. Daran, 1989). The spectrum reveals a large number of very closely spaced spectral lines which are not always resolved. Some of the bandheads seen in trace (a) can be identified easily but other lines are more difficult to assign. This is due to the fact that the laser induces also transitions between high-lying states other than the A and B states. On one hand this is a complication for the interpretation of the spectra, on the other hand it immediately shows the capability of OG spectroscopy to detect transitions which may otherwise be hidden in relation to much stronger features when other spectroscopic detection methods are used.

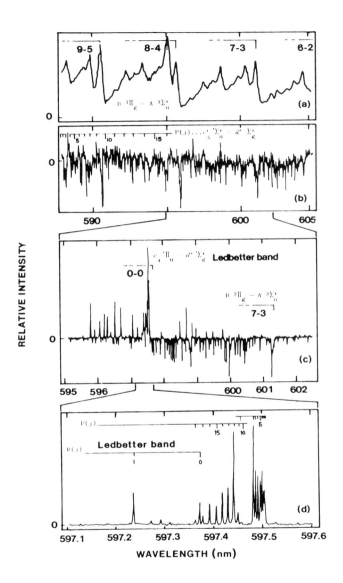

Figure 6: Spectra of N_2 in a hollow–cathode discharge; (a) emission spectrum; (b) OG spectrum using a pulsed laser with $\Delta \nu \simeq 3$ GHz (after Daran, 1989); (c) OG spectrum using a broadband cw laser with $\Delta \nu \simeq 2$ GHz (after Pfaff et al, 1984); (d) OG spectrum using a single–mode cw laser with $\Delta \nu \simeq 1$ MHz (after Suzuki and Kakimoto, 1982). Some prominent transition bands are indicated; for details see text.

As an example, for the c_5' $^1\Sigma_u^+ \rightarrow$ a" $^1\Sigma_g^+$ transition of N_2 part of a P–branch is assigned (the scale of the figure does not readily reveal this).

When a cw laser is used which in general has a better resolution than a pulsed dye laser it is seen that line narrowing to single mode and sub– Doppler techniques may be necessary to fully reveal the rotational structure in the spectra. In trace (c) a cw laser of moderate bandwidth is used ($\Delta\nu \sim 2$ GHz); the resolution is hardly better than for a pulsed laser although the signal–to–noise ratio is improved because of the higher duty cycle of the system. Again, some of the spectral features are indicated in the figure. It is interesting to note that the c_4 $^1\Pi_u \rightarrow$ a" $^1\Sigma_g^+$ band has different polarity in traces (b) and (c). This is due to the fact that the discharge conditions are not the same, and that the signal recovery is by gated technique and phase–sensitive detection, respectively. The temporal shape of the signals for the various transitions is different due to diferent relaxation times of the levels involved in the transition. Signal pulses often have both positive and negative components; these may be at different times in the pulse for different transitions, and thus it depends on the setting of the integrator gate position if a positive or negative average is recorded. For a fuller discussion of this see Pfaff et al (1984) and Daran (1989).

When a single mode dye laser ($\Delta\nu \simeq 1$ MHz) is used it becomes possible to largely resolve the full rotational structure, as is shown in trace (d) for the Ledbetter band of N_2. But even then not all rotational lines are completely separated, as can be seen at the position of the head of the Q–branch at around 597.5 nm, and sub–Doppler OG techniques have to be applied (see Suzuki and Kakimoto, 1982).

Investigation of molecular fragments

The existence of a number of radicals and reaction intermediates has been postulated to be of importance for a number of combustion and discharge processes; however, spectroscopic studies are often difficult to perform since not sufficiently high concentrations of the species can be obtained. In this respect discharges form a rather elegant tool to produce molecular fragments since the electron bombardment of the parent molecule often excites states which are highly predissociative thus producing the desired fragment.

An example is the radical HCO which had been postulated to be an important constituent in hydrocarbon–oxygen flames but its spectroscopy, necessary for the understanding of its formation and consumption, was rather incomplete. This is largely due to the fact that many of the well established laser techniques cannot be used; for example, the normally powerful method of LIF fails in the study of the lowest excited state of HCO since its fluorescence quantum yield is extremely small.

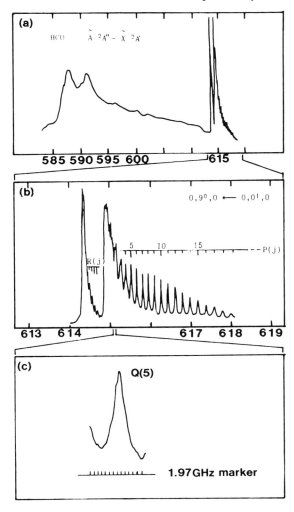

Figure 7: OG spectrum of HCO $(\tilde{A} - \tilde{X})$, produced in a RF discharge of acet–aldehyde; (a) and (b) low–resolution spectra, laser bandwidth $\Delta\nu \simeq 30$ GHz and (c) high–resolution spectrum, laser bandwidth $\Delta\nu \simeq 1$ MHz (after Vasudev and Zare, 1982).

Exploiting the OG effect accompanying laser excitation in a discharge or flame such problems may be largely overcome, and concentrations of the parent molecule necessary for the detection of a particular radical are expected to be orders of magnitudes smaller than those needed in other approaches.

High resolution OG spectroscopy of HCO is described by Vasudev and Zare (1982). Briefly, the HCO radical is produced in an RF discharge through acetaldehyde, CH_3CHO, at a pressure of ~ 2 mbar. Excitation in the $\tilde{A}-\tilde{X}$ transition is achieved using a cw dye laser in multimode ($\Delta \nu$ ~ 30 GHz) or single mode ($\Delta \nu$ ~ 1 MHz) operation. Examples for low and high resolution spectra are shown in Figure 7. It was not only possible to produce very precise molecular data from the spectra but also to gain insight into level populations, dynamics and predissociation rates. That it is possible to extract such a wealth of information and the ease with which such experiments can be performed suggests that the OG technique may also be used in the study of other (nonradiative) processes for species which can be produced only in low concentrations.

Investigation of reaction products

Similar to the case of reactive flames the plasma conditions in a discharge may be favourable for the formation of reaction products from the constituents in the plasma. In particular metal complexes lend themselves for such investigations since pure metals or conductive (or semiconductive) salts can be used to form the inner surface of a hollow cathode; the metal is then introduced into the glow discharge by sputtering.

An example of this procedure is given by Carlson et al (1985) who investigated TiO molecules. The molecules are formed in the interaction of atomic titanium, Ti, and titanium dioxide, TiO_2, sputtered from the inner surface of a hollow cathode composed of a highly compressed mixture of these two substances; argon was used as the buffer gas.

A typical OG spectrum for TiO is shown in Figure 8 which also includes recordings of a TiO emission spectrum (trace a) and an OG spectrum of argon (trace c) for comparison. The complex sub-band structure of the investigated B $^3\Pi$ – X $^3\Delta$ system leads to strong overlap in the spectrum, and thus resolution of the rotational structure is very difficult to disentangle. Also, the reaction product is produced in only minor concentrations, and some of the molecular features are swamped by the much stronger OG signals from transitions in the argon buffer gas. Nevertheless, it is even possible to deduce from the intensity of the observed sub–bands reaction and excitation dynamics of TiO which is further discussed in Carlson's paper.

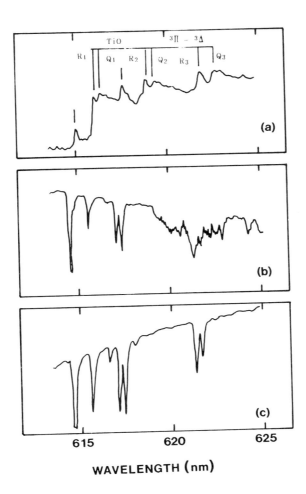

Figure 8: Spectra obtained from a Ti/TiO$_2$ hollow–cathode discharge; (a) low–resolution emission spectrum of TiO; (b) OG spectrum of TiO in argon buffer gas; (c) reference OG spectrum of pure argon (after Carlson et al, 1985).

Investigation of collision complexes

At low pressure, often the only force noticeable between two atoms is the van der Waals' force, apart from the repulsive forces which come into play at extremely small internuclear distances. Such van der Waals' molecules exhibit only very shallow potential wells at large internuclear separations, and even at low temperature the kinetic energy of two colliding atoms is large enough to prohibit them from being "trapped" in the potential minimum to become a molecule.

However, complexes being only extremely weakly bound in their ground state may exhibit relatively deep potential minima in their excited states to form well behaved molecules accessible to spectroscopy. However, they often can only be formed in very low concentrations. In the majority of cases they are characterized from spectroscopic data gained from absorption processes during atomic collisions, and signals are very weak indeed. In particular, rare gas dimers, metal atom – rare gas and halogen atom – rare gas systems have found much attention because of their importance in, for example, broad–band lamps and excimer lasers. However, in general, only data for transitions between the ground state and the lowest excited states are available from normal low–pressure neutral–gas collision experiments whereas transitions between highly excited states often evade observation. This can be different in a discharge environment.

An example is the homonuclear molecule He_2. In its ground state it only exhibits a van der Waals minimum; on the other hand, some of the excited states of He_2 have rather deep wells. Excited He_2 states can be generated for example in discharges at elevated pressures and high current densities. Significant population in Rydberg states of He_2 in a discharge plasma can be obtained via two major channels:

(i) Three–body He metastable recombination
$$He^* + 2 \cdot He(1s) \rightarrow He_2^* + He(1s) + \Delta E$$
(ii) Recombination of He_2^+ molecular ions
$$He_2^+ + e^- + M \rightarrow He_2^* + M + \Delta E$$

M can be another electron or a He(1s) ground state atom; for a wider discussion of He_2^* production, as a function of pressure and discharge current, see e.g. Ramalingam (1990).

Molecules 17

An example of a molecular He$_2$ OG spectrum is shown in Figure 9 for the (0–0) band of the f $^3\Pi_u$ – b $^3\Pi_g$ transition; R– and P– lines are clearly identified. Again, as outlined earlier for N$_2$, some of the signals have opposite sign for pulsed (trace a) and cw (trace b) excitation. Also, the population within the individual rotational levels is rather different since the operation conditions under which the spectra were obtained vary

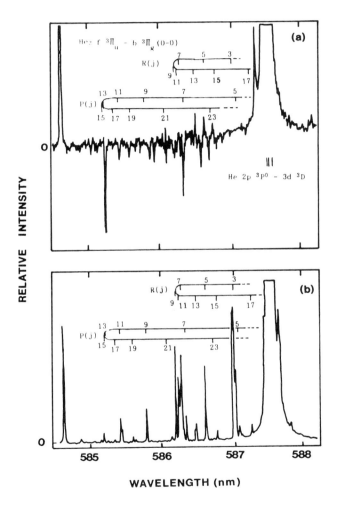

Figure 9: OG spectra of He$_2$ in a hollow–cathode discharge; (a) at 15 mbar of helium and a current of 50 mA, using a pulsed laser with $\Delta v \simeq 3$ GHz (after Daran, 1989); (b) at 13 mbar of helium and a current of 300 mA, using a cw laser with $\Delta v \simeq 2$ GHz (after Pfaff et al, 1984).

significantly. Besides the well resolved rotational structure of the f–b transition strong asymmetry in the atomic He peak near 587.5 nm is observed. This "wing" to the atomic line changes with gas pressure and discharge current, and is typical for van der Waals type complexes; the spectra of He_2 complexes will be discussed in more detail in Telle et al (1990a).

Another example is the metal–atom rare–gas complex CaNe. In Figure 10 part of an OG spectrum is shown which was obtained from a hollow cathode discharge in a Ca/Na/Ne mixture. In the enlarged part of the spectrum a very asymmetric profile is observed for the Ca line at 585.7 nm. The linewidth of the pulsed laser used in this experiment was of the order of 3 GHz (or 0.036 nm), much narrower than the observed width of the line profile of around 0.3 nm. From the operating

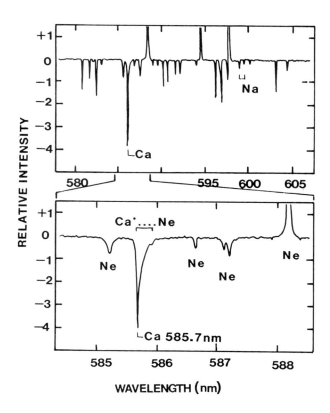

Figure 10: OG spectrum of a Ca/Na/Ne hollow–cathode discharge; the Ca transition 4s4p $^1P_1$ – 4p $^1D_2$ at 585.7 nm exhibits an asymmetric line profile ascribed to the formation of a Ca*··Ne complex.

parameters within the discharge (total pressure ~ 5 mbar, current ~ 25 mA) and rare gas reference lines it is estimated that contributions from Doppler broadening (~ 0.5 GHz) and pressure broadening (~ 0.25 GHz) are very much smaller than the width of the observed line profile. Thus it can be concluded that the signal has to be ascribed to a transition in an excited collisional complex of Ca*–Ne; the potentials connected in the transition correlate asymptotically to the excited atomic states Ca(4s4p $^1P_1$) and Ca(4p $^1D_2$); further discussions for this and other transitions will be found in Telle et al (1990a).

These measurements of excited collision complexes is further evidence for the power of OG spectroscopy, as has been pointed out already earlier, when a species is under investigation which can be produced in only minute concentration.

Investigation of molecular ions

The detection and spectroscopy of molecular ions is a subject of great interest because of their importance in for example plasmas, processes in the upper atmosphere or laser chemistry. OG spectroscopy in electrical discharges is a valuable tool since the discharge can be both a source and a convenient detector. While a substantial number of neutral molecules has been investigated exploiting the optogalvanic effect to our knowledge the only molecular ions which have been detected using OG spectroscopy are N_2^+ and CO^+.

One of the main differences in the OG effect for neutrals and ions is the following. Laser–induced transitions in neutrals alter the probability for ionization by disturbing the steady–state population distribution thus inducing an impedance change in the discharge. For ions the impedance change has a different origin involving direct modification in ion mobility; the change in mobility is due to a difference in collisional properties of ground–state ions versus excited–state ions. The validity of this mechanism can be verified when studying the spatial, temporal and pressure dependence of the OG signal.

An illustrative example is given by Walkup et al (1983) who investigated N_2^+, in particular a number of rotational lines in the (0,0) band of the $X\ ^2\Sigma_g^+ \rightarrow B\ ^2\Sigma_u^+$ transition. Pure nitrogen at a pressure of ~ 1.3 mbar was discharged between plane parallel electrodes at a distance

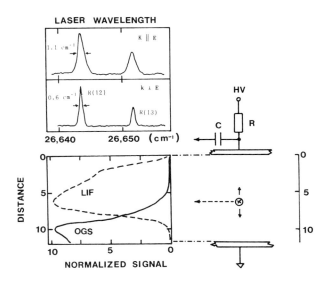

Figure 11: Spatial dependence of optogalvanic signals (OGS) and laser– induced fluorescence signals (LIF) in a glow discharge of N_2 at 1.3 mbar. The laser excitation is in the $X\ ^2\Sigma_g^+ - B\ ^2\Sigma_u^+$ transition of the N_2^+ molecular ion. The lineshapes for excitation perpendicular or parallel to the electric field reveal line broadening due to the ion motion towards the cathode (after Walkup et al, 1983).

supporting abnormal glow (see left–hand side of Figure 11); typically the DC discharge current was kept at around 7 mA. The OG signal induced by the pulsed tunable dye laser is monitored through a DC blocking capacitor and processed with a boxcar integrator. The upper state of the transition can also be observed via radiative decay in monitoring the $B(v'=0) \to X(v''=1)$ LIF signal following excitation in the (0,0)–band. When monitoring the intensity of the OG and LIF signals as a function of position in the discharge the spatial dependence of the two is found to be sharply different (see Figure 11). While the OG signal is only substantially strong for excitation within the cathode sheath region of the discharge the opposite is true for the LIF signal which has its maximum in the negative glow region. Since the LIF signal is approximately proportional to the ground state N_2^+ density it shows that the N_2^+ density is substantially larger in the glow region than in the cathode region. The spatial sensitivity of the two detection methods thus complements each other.

Beyond the relative level populations (from the spectral intensity of individual rotational transitions) and the spatial distribution of the molecular ion density it is also possible to use OG detection as a diagnostic tool to obtain ion kinetic energies. In an ordinary gas the linewidth of a transition is usually dominated by Doppler broadening due to the kinetic energy distribution of the particles. Ions in a DC field exhibit an additional Doppler shift parallel to the field due to the accelerated ion motion towards the cathode. This will become evident when line profiles of a transition are scanned using a narrow-band tunable laser, comparing the signals generated when the laser beam propagates perpendicular and parallel to the electric field. An example is given in the upper part of Figure 11 for two rotational lines; the difference in linewidth is quite evident, and the contribution of the field to the ion motion can be deconvoluted from the line profile.

The technique thus far has been used for molecular ions in their ground state exploiting the fact of different ion mobility, and hence altered collisional properties, on laser excitation. How this effect may be exploited for transitions between higher excited states is not immediately obvious; in particular one may be faced with the difficulty that excited state population densities of molecular ions are extremely small due to their very high potential energy. Also, mobility changes for ions in short-lived excited states could be rather small thus preventing their detection. However, there are specific cases where relatively large densities can be generated. In particular this has been observed for the formation of $(He\cdot\cdot Cd)^+$ molecular ion complexes formed in a cw hollow cathode $HeCd^+$ metal vapour laser; the population in the molecular ion state was even sufficiently high to exhibit laser action. Currently gain measurements for a number of molecular $HeCd^+$ ion bands are under way exploiting the OG effect when seeding the laser tube with radiation from an external tunable laser (Telle et al, 1990b)

OPTOGALVANIC DETECTION OF PHOTODISSOCIATION

When irradiating vapours of metal halides with light in the UV or VUV spectral range dissociation takes place and often pairs of positively charged metal ions and negatively charged halide ions are formed. If this process is initiated between electrodes an applied field accelerates the positive ions towards the cathode and the negative ions towards the anode, and the generated current gives rise to an optogalvanic type signal. The first experiments of OG spectroscopy of metal halides were performed by Terenin (1930). In his and most later experiments a major problem has been corrosion of the electrodes due to chemical reactions with the hot halide compounds. This problem can be overcome when the electrodes are placed outside the reaction volume thus avoiding contact with the vapour; this has recently been demonstrated by Schnell and Luthy (1985) for photoionization of thallium iodide, TlI.

The experimental configuration is schematically shown in Figure 12 (left-hand side). The metal halide vapour is contained in a small quartz cell of approximately 1 cm^2 cross section which is heated to provide sufficiently large particle densities ($\sim 10^{15}$ cm^{-3}). The cell, together with the external electrodes, forms a capacitor which for the particle densities in these experiments had a capacitance of $C_c \simeq 50$ fF. Photodissociation of TlI into ion pairs produces changes in the capacitance of the cell and hence an OG signal can be detected in measuring the induced voltage drop across the load resistor R_1.

The OG signal due to ion pair production in thallium iodide vapour, as a function of excitation wavelength, is shown in Figure 12 (right-hand side). Standard deconvolution procedures can be used to extract information on the molecular potentials from the dissociation spectrum.

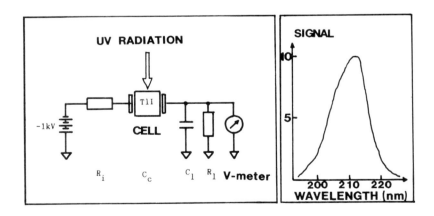

Figure 12: OG spectroscopy of photodissociation with external electrodes (impedance spectroscopy); (a) experimental set-up – $R_i \simeq 1$ MΩ charging resistor, $C_c \simeq 50$ fF cell capacitance, $C_1 \simeq 200$ pF damping capacitor, $R_1 \simeq 100$ MΩ load resistor; (b) ion pair production in thallium iodide vapour, as a function of dissociation wavelength (after Schnell and Luthy, 1985).

The method is not only applicable to metal halides but in principle to all molecular species for which ion pairs are formed, i.e.

$$AB + (n)\cdot h\nu \rightarrow A^+ + B^-$$

where dissociation is achieved by single- or multi-photon excitation.

The OG detection of photodissociation is not necessarily restricted to cases in which ion pairs are formed. Equally suitable are processes in which the molecule dissociates into neutrals followed by consecutive ionization of one of the products, resulting for example in a sequence

$$AB + (n)\cdot h\nu \rightarrow (AB)^* \rightarrow A\ (A^*) + B^*\ (B)$$

followed by one or more processes of the type

$$A\ (A^*) + (n)\cdot h\nu \rightarrow A^+ + e^-$$

$$A\ (A^*) + h\nu + A \rightarrow (A^{**}\cdots A) \begin{array}{l} \rightarrow A_2^+ + e^- \\ \rightarrow A^+ + A^- \end{array}.$$

An example for this scheme has been reported by Naqvi (1985) for two-photon dissociation of Tl_2.

SUMMARY

In this review an attempt has been made to establish the usefulness of OG spectroscopy for the investigation of molecular species. We hope that in particular it has become obvious that it is not only possible to detect molecules utilizing the OG effect but that for a number of applications it may be the only means to have access to the species under investigation. This is in general true for highly excited molecular states, the investigation of molecular fragments and radicals, and excited collision complexes, all of which may only be accessible in a plasma environment. Readily available laser sources allow spectroscopy at medium and high resolution thus revealing rotational structures; even sub-Doppler resolution is obtainable opening up the possibility to gain insight into dynamics of the species in a plasma environment and thus using OG spectroscopy as a plasma diagnostic tool. Although most efforts in OG spectroscopy have been directed towards the investigation of atomic species the list of molecules is ever growing.

STABLE GASEOUS MOLECULES	H_2 / I_2 / N_2 / CO
	CO_2 / NO_2 / NH_3 / $C_2H_2O_2$

FRAGMENTS AND REACTION PRODUCTS	
molecular gas parents	NO / CN / HCO / NH_2
	HCO_2 / H_2CO / HNO_3
flame reactions	ScO / YO / LaO / BaCl
	SrOH / BaOH
discharge reactions	TiO / CuO
photodissociation	Tl–, Ga–, In–, Na– halides
	Tl_2

COMPLEXES	
rare gas dimers	He_2
metal dimers	Li_2 / Cs_2 / Rb_2
rare gas – metal atom	CsKr / CsAr / Cs_2Kr / CaNe / NaNe

MOLECULAR IONS	N_2^+ / CO^+ / $HeCd^+$

A collection of many of the molecules detected via OG methods is given in the table.

We anticipate that in the future more species will be investigated if the need arises. The simplicity with which the technique can be applied seems to be very intriguing; on the other hand, the interpretation of complex spectra of usually highly excited states may be rather difficult since often little spectroscopic information is available for these states which could be used as a starting point in the evaluation procedure. However, in particular for complexes and radicals, no other means than OG detection in a controlled plasma environment may be able to unravel features of those species for high–lying states.

REFERENCES

Ausschnitt C P, Bjorklund G C and Freeman R R 1978 Appl.Phys.Lett. **33** 851

Carlson R C, Cross L A and Dunn T M 1985 Chem.Phys.Lett. **113** 515

Carswell A I and Wood J I 1967 J.Appl.Phys. **38** 3028

Collins C B, Johnson B W, Popescu D, Musa G, Pascu M L and Popescu I 1973 Phys.Rev.A **8** 2197

Collins C B, Curry S M, Johnson B W, Mirza M Y, Chellehmalzadeh M A, Anderson J A, Popescu D and Popescu I 1976 Phys.Rev.A **14** 1662

Daran A B M 1989 Ph.D. thesis University College of Swansea

Erez G, Lavi S and Miron E 1979 IEEE J.Quant.Electron. **15** 1328

Ernst K and Inguscio M 1988 Revista Nuovo Cimento **11** series 3 number 2

Feldmann D 1979 Opt.Commun. **29** 67

Foote P D and Mohler F L 1925 Phys.Rev. **26** 195

Goldsmith J E M and Lawler J E 1981 Contemp.Phys. **22** 235, and references therein

Miron E, Smilanski I, Liran J, Lavi S and Erez G 1979 IEEE J.Quant.Electron. **15** 194

Naqvi A S 1985 J.Chem.Phys. **82** 2217

Penning F M 1928 Physica **8** 137

Pfaff J, Begemann M H and Saykally R J 1984 Mol.Phys. **52** 541

Ramalingam P 1990 Ph.D. thesis University College of Swansea

Schenck P K, Mallard W G Travis J C and Smyth K C 1978 J.Chem.Phys. **69** 5147

Schnell S and Luthy W 1985 J.Phys.E:Sci.Instrum. **18** 28

Skolnick M L 1970 IEEE J.Quant.Electron. **6** 139

Suzuki T and Kakimoto M 1982 J.Mol.Spectrosc. **93** 423

Telle H H, Daran A B M and Karyono 1990a to be published in Chem.Phys.

Telle H H, Karyono and Hassan M 1990b accepted for publication in Laser Chemistry

Terenin A 1930 Phys.Rev. **36** 147

Turk G C, Mallard W G, Schenck P K and Smyth K C 1979 Analyt.Chem. **51** 2408
  and references therein

Vasudev R and Zare R N 1982 J.Chem.Phys. **76** 5267

Walkup R, Dreyfus R W and Avouris Ph 1983 Phys.Rev.Lett. **50** 1846

Webster C R and Rettner C T 1983 Laser Focus **19** 41, and references therein

Webster C R and Menzies R T 1983 J.Chem.Phys. **78** 2121

Laser studies of discharge plasma sheaths

J. E. Lawler and E. A. Den Hartog

Department of Physics, University of Wisconsin, Madison, WI 53706 USA

ABSTRACT: Laser optogalvanic and fluorescence diagnostics are used to study the ionization and power balance of the negative glow region in a cold cathode He discharge. Monte Carlo simulations, based on accurate field maps and normalized to a current balance at the cathode from the field maps, are used to study the hot electrons in the negative glow. The electric field in the cathode fall region is mapped using optogalvanic detection of Rydberg atoms. Laser diagnostics are used to measure the density and temperature of the cold electrons in the negative glow. These cold electrons are trapped in a potential energy well. The ionization balance of the negative glow is dominated by a process analogous to ambipolar diffusion in which the ions drift in a reversed field to the anode. The power balance of the cold trapped electrons is dominated by cooling due to elastic collisions on neutrals, and heating due to Coulomb collisions with hot electrons. Most of the light from the negative glow is due to the hot electrons exciting atoms to radiating levels from metastable levels.

1. INTRODUCTION

The cathode region of glow discharge plasmas is of fundamental and practical interest. The proximity of the cathode and the large and rapidly varying electric fields result in electron energy distributions which are not determined by the local E/N (ratio of electric field to gas density). The failure of this widely used hydrodynamic equilibrium (or local field) approximation means that a realistic model should be based on kinetic theory. Practical interest arises from the many important uses of glow discharge plasmas.

Significant progress has been made in recent years in understanding the high field or cathode fall part of the cathode region. The development of laser spectroscopic techniques for mapping electric fields has advanced our understanding of the cathode fall. Doughty et al.(1984a,b)

© 1991 IOP Publishing Ltd

developed a technique for field measurements based on optogalvanic detection of Rydberg atoms. Moore et al.(1984) developed a laser induced fluorescence technique for field measurements. Accurate field maps generated with these noninvasive techniques have been used in many studies. For example, Doughty et al.(1987) were able to extract the current balance (ratio of ion to electron current) at the surface of a cold cathode from accurate field maps and gas density measurements. Recent theoretical developments have also greatly improved our ability to model the avalanche which occurs in the cathode fall region when an electron is emitted from the cathode. The null collision Monte Carlo technique for nonuniform fields developed by Boeuf and Marode (1982) has been very valuable. The Convective Scheme has enabled Sommerer et al.(1989a) to do fully self-consistent kinetic simulations of the cathode fall region.

The negative glow is the part of the cathode region immediately next to the high field cathode fall. Although it has a very small electric field, it too is nonhydrodynamic because of a low density of high energy electrons injected from the cathode fall region. Den Hartog et al.(1989) recently described laser experiments which demonstrated the existence of quite high densities ($n_e^c = 5\times10^{11}$ cm^{-3}) of rather low energy electrons ($k_B T_e^c = 0.12$ eV) in the negative glow of a He glow discharge. The electron temperature T_e^c and density n_e^c refers to the "cold" negative glow electrons. The high energy electrons injected from the cathode fall result in a distribution which is only partly described by a single Maxwellian. These laser experiments detected mixing of populations between excited levels of He caused by collisions between low energy electrons and excited He atoms. Data from one exothermic (superelastic) and one endothermic (inelastic) reaction were combined to yield a density and temperature of the low energy electrons.

The low energy electrons are trapped in a potential energy well formed by a field reversal in the negative glow. There has been much discussion in review articles on the existence and location of a field reversal in the negative glow (Francis 1956, Long 1978, Emeleus 1981). Most authors seem comfortable with a field reversal in the negative glow under a variety of conditions. A convincing theoretical treatment of the cathode fall and negative glow regions which predicts the existence and location of the field reversal and other properties of the negative glow has been an elusive goal. A convincing theoretical treatment must avoid the many assumptions and approximation of older models.

It is difficult to directly measure the very small fields in the negative glow using either probe or spectroscopic techniques. In an important early work Druyvesteyn (1937) produced evidence for the field reversal by measuring the discharge voltage at fixed current while moving the anode into and through the negative glow. Druyvesteyn's measurements, like more modern probe measurements, are rather invasive and thus perturb the discharge. Den Hartog et al. (1988) presented evidence for a field reversal in the negative glow of dc He discharges. Their evidence is based on a comparison of Monte Carlo simulations of electron avalanches with an empirical current balance at the cathode. Very few of the ions produced in the negative glow contribute to the ion current at the cathode in discharges studied by Den Hartog et al.(1988). The ionization in the negative glow produced by energetic electrons injected from the cathode fall must be balanced either by recombination or, if recombination is negligible, then by a drift of ions in a reversed field toward the anode. This drift is analogous to ambipolar diffusion. More recently Gottscho et al.(1989) have observed a change in the sign of an optogalvanic effect on molecular ion lines. The sign change was correctly interpreted as due to a field reversal in the negative glow. The noninvasive optogalvanic technique for observing the field reversal is attractive because probes can represent a significant perturbation to the plasma, and because probe characteristics are difficult to interpret for non-Maxwellian, anisotropic electron energy distributions.

A convincing theoretical treatment of the ionization balance, which would predict the field reversal and the density of the cold electrons, requires a self consistent kinetic calculation. The Convective Scheme (Sommerer et al. 1981a, 1989b) shows great promise of making such calculations possible for a variety of discharges in the near future. The research described here focuses on identifying important physical processes in the negative glow and on some benchmark experiments for comparison to models. Section 2 is a detailed discussion of the laser experiments by Den Hartog et al.(1988,1989) which mapped the fields in the cathode fall and determined the density and temperature of the cold negative glow electrons in He discharges. Section 3 describes Monte Carlo simulations of electron avalanches in the same discharge. These simulations are used to study the hot electrons in the negative glow. The simulations are quite realistic because they are based on empirical field maps, gas density measurements, and an excellent set of He cross sections. Section 4 describes the previously mentioned model of the ionization balance in the negative glow presented by Den Hartog et al.(1988). Section 5 explores the power balance of the cold electrons using both experimental results and Monte Carlo simulations. This work identifies the dominant physical processes which maintain $k_B T_e^c$ between 0.1 eV and 0.2 eV. We find that the dominant cooling term in the power balance of the cold electrons is due to energy lost during elastic collisions with neutral atoms. We find that the dominant heating term in the power balance of the cold electrons is due to energy gain in Coulomb collisions with the hot negative glow electrons. Section 6 explores the role of multistep excitation caused by hot electrons in producing light from the negative glow. Finally Section 7 summarizes our conclusions and outlines future work.

2. EXPERIMENTS

2.1 Experimental Apparatus

The experiments described in this paper were performed in a clean, reproducible, nearly one dimensional He discharge. Helium was chosen because it is ineffective at sputtering the cathode, and because the cross section data base is better for He than for any other gas.

Figure 1 is a schematic of the experimental apparatus. The discharge used in these studies is produced between flat circular Al electrodes 3.2 cm in diameter and separated by 0.62 cm. The electrodes are water-cooled to minimize gas heating. The discharge tube is made primarily of glass and stainless steel. Most of the large seals are made with knife edge flanges on Cu gaskets. The only exceptions are the high vacuum epoxy seals around the fused silica Brewster windows. A liquid N_2 trapped diffusion pump evacuates the tube to 2×10^{-8} Torr. When no liquid N_2 is in the trap an ion pump maintains the vacuum to prevent oil from back-diffusing into the system. The leak rate into the sealed discharge tube is approximately 3×10^{-4} Torr/day. For discharge operation ultra high purity (0.999999) He is slowly flowed through the system. A capacitive manometer monitors the pressure which is maintained at 3.5 Torr. The He first passes through a cataphoresis discharge to remove any residual contaminants before entering the main discharge tube. Emission spectra reveal only weak Al and H impurity lines. Although He is much less effective at sputtering than heavy inert gases, some sputtering of the Al cathode does occur. We suspect that the slow erosion of the cathode gives rise to both the weak Al and H impurity lines, since many metals absorb hydrogen. The surface of the Al cathode is cleaned in-situ by running an Ar discharge. Argon is very effective at sputtering most metals including Al. A carefully prepared cathode provides discharge voltage-current characteristics which are stable to a few percent over a period of a month.

Experiments are carried out over a range of discharge current densities from 0.190 mA/cm^2 to 1.50 mA/cm^2. The low current density corresponds to a near normal cathode fall voltage of 173 V and the high current density corresponds to a highly abnormal cathode fall voltage of 600 V. The discharge current is spread uniformly across the surface of the electrodes. The absence of significant fringing was verified by segmenting the 3.2 cm diameter cathode into a 1.6 cm diameter disc and a close fitting annulus with an outside diameter of 3.2 cm. Each part of the cathode was maintained at the same potential during operation and the average current density on each was measured. The current density was found to be uniform across the cathode. This measurement is one of the justifications for using one dimensional models of the cathode fall region. The segmented cathode was replaced by a solid cathode after the current density measurements.

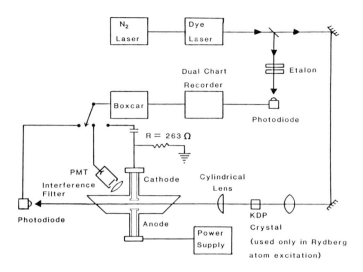

Fig. 1. Schematic of the experimental apparatus showing three detection methods: optogalvanic, fluorescence, and absorption detection.

The laser used is a N_2 laser pumped dye laser. Several different configurations for the dye laser are used, depending on the specific experiment. The dye laser bandwidth is 0.3 cm^{-1} without an etalon. An etalon is used to reduce the bandwidth to 0.01 cm^{-1} (300 MHz) for some measurements. The dye laser with frequency doubling is tunable over a wavelength range 200-700 nm.

Figure 1 shows all three detection schemes used in these experiments. Optogalvanic detection, fluorescence detection, and absorption detection each have unique advantages which have been discussed (Den Hartog et al. 1988). In order to perform spatially resolved measurements without disturbing laser alignment the discharge is mounted on a precision translation stage. Good spatial resolution is achieved, when necessary, by passing the laser beam through a narrow slit and imaging the slit into the discharge region.

2.2 Electric Field Maps and the Current Balance

Accurate empirical maps of the electric field in the cathode fall region are used in Monte Carlo simulations of electron avalanches in the cathode fall and negative glow regions. Although spectroscopic techniques are not sufficiently sensitive to directly map the electric field in the negative glow, the maps of the cathode fall field insure that the Monte Carlo simulations are as realistic as possible. The behavior of the energetic or "hot" electrons is insensitive to the small negative glow fields; it is determined by the large cathode fall fields. The field maps in the cathode fall region also determine the current balance at the cathode surface which provides a normalization for the Monte Carlo simulations. The Monte Carlo simulations then provide the density and average energy of the hot electron in the negative glow, the absolute ionization rate per unit volume in the negative glow, and important insight on the ionization and power balance in the negative glow. Monte Carlo techniques are not suited for studying the cold trapped electrons.

The determination of the current balance (Doughty et al. 1987) involves a straightforward analysis of the field maps, but it does require quite accurate field maps. A spatial gradient of the electric field determines the ion charge density near the cathode by Poisson's equation. The electron density is negligible in the high field cathode fall region. This analysis depends on a rigorous proof that the ions are in hydrodynamic equilibrium throughout most of the cathode fall region. The E/N near the cathode surface determines the ion drift velocity because the ions are in hydrodynamic equilibrium. An ion current density determined as the product of the ion charge density and ion drift velocity is subtracted from the total discharge current density to determine the electron current density.

Optogalvanic detection of Rydberg atoms (Doughty et al. 1984a) is used to map the electric field in the cathode fall region. Fragile Rydberg atoms are an ideal discharge probe. They exhibit large, linear Stark effects which can be accurately (~1%) measured, and easily interpreted using nearly hydrogenic single-electron calculations. Optogalvanic methods make it possible to detect the Rydberg atoms which do not survive long enough to fluoresce in a typical discharge. Laser techniques have a number of important advantages over the traditional electron beam deflection technique (Warren 1955). Laser techniques are useful at higher pressures and discharge current densities. It is easier to use a spectroscopic technique in a high purity discharge system. A laser technique is truly nonperturbing when performed with a nanosecond pulsed laser because the field is measured when the atoms or molecules absorb the laser light, and any perturbation to the discharge fields occurs on a longer time scale. Lasers have the potential for making measurements with a few microns spatial resolution and nanosecond temporal resolution. A technique based on Rydberg atoms has broad applicability because all atoms and molecules have Rydberg levels. It also has a wide dynamic range because one can choose an appropriate principal quantum number to achieve a desired level of sensitivity to the electric field.

A much more detailed discussion of electric field measurements using optogalvanic detection of Rydberg atoms has been published (Den Hartog et al. 1988). Figure 2 shows the electric field maps for five different current densities. The fields are accurate on average to ~1%. This accuracy is verified by integrating the fields across the cathode fall to determine a potential difference within ~1% of the discharge voltage

Plasma Sheaths

measured using a digital voltmeter. The field magnitude decreases almost linearly with distance from the cathode. A slight negative curvature is visible in Fig. 2. The spatial gradient of the field gives the ion charge density according to Poisson's equation.

The average ion velocity can be determined from the field maps if the ions are in hydrodynamic equilibrium and if the gas density is known. Ion motion is limited by the symmetric charge exchange reaction

$$He^+(fast) + He(slow) \rightarrow He(fast) + He^+(slow). \tag{1}$$

This reaction has a large cross section which is only very weakly dependent on energy. Charge exchange effectively stops an ion every collision and wipes out all memory of its history. It seems intuitive that ions, unlike the electrons, should be in hydrodynamic equilibrium

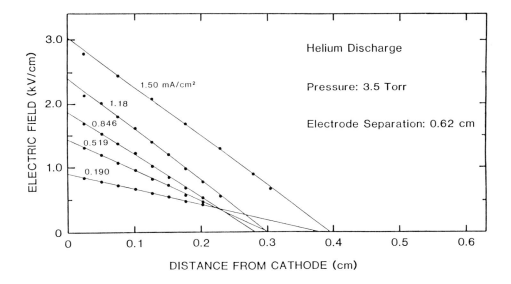

Fig. 2. Electric field as a function of distance from the cathode for five current densities, all at 3.50 Torr. The lines are linear-least-square fits to the data. The anode corresponds to the right-hand side of the figure.

throughout most of the cathode fall region, because this region has a thickness equivalent to 50 → 100 mean-free-paths for symmetric-charge-exchange. The large field gradient, and the distributed source of new ions from electron impact ionization throughout the cathode fall, makes one reluctant to rely on an intuitive "proof" that ions are equilibrated. Lawler (1985) has published a rigorous proof based on analytic solutions to the Boltzmann equation for ions. In this work exact Green's functions for the Boltzmann equation were used to compute ion equilibration distances for four idealized situations. Equilibration distance is here defined as the distance required for the average velocity of the ions to approach within 10% of the equilibrium drift velocity. The four idealized situations and corresponding equilibration distances (expressed in mean-free-paths) are: A) a planar source of ionization in a uniform field (0.65), B) a uniform source of ionization in a uniform field (4.5), C) a planar source of ionization near the origin of a linearly increasing field (1.7), and D) a uniform source of ionization in a linearly increasing field (5.7). Numerous Monte Carlo simulations of electron avalanches have shown that the ionization rate per unit volume peaks near the cathode fall-negative glow boundary where the electric field extrapolates to zero (Segur et al. 1983, Doughty et al. 1987, Den Hartog et al. 1988). Thus cases C) and D) provide a rigorous lower and upper bound for the ion equilibration distance as measured from the cathode fall-negative glow boundary. The ions are in hydrodynamic equilibrium throughout the cathode fall, except for a narrow region within 6 mean-free-paths of the cathode fall-negative glow boundary.

The final requirement for determining the current balance is a measurement of the gas density in the cathode fall. The discharge tube pressure in these experiments was fixed at 3.5 Torr. In the cathode fall region of an abnormal discharge gas heating is often significant. The symmetric charge exchange reaction which produces a short ion equilibration distance also serves to convert most of the electrical power dissipated in the cathode fall directly into heavy particle (atomic) motion and into heat. Doughty et al. (1987) used the Doppler width of a carefully selected He transition to determine the gas temperature and gas density versus discharge power shown in Fig. 3. The linewidth of the 501.6 nm transition from the 2^1S to 3^1P level was measured using laser optogalvanic detection. Various contributions to the observed linewidth including a laser bandwidth of 300 MHz, a natural width of 92 MHz, pressure broadening at 3.5 Torr of 146 MHz, and Stark broadening on the order of 100 MHz, are much smaller than the Doppler width at 293K of 3.66 GHz. It was somewhat surprising that the water cooled electrodes were not more effective in limiting the gas temperature rise, but the electrode assemblies have large thermal impedances. The gas temperature rise is consistent with the expansion of the cathode fall region at high current densities as shown in Fig. 2.

Plasma Sheaths

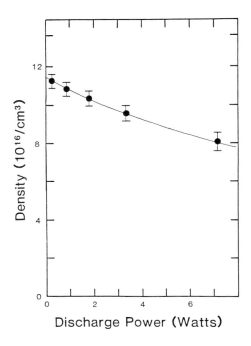

Fig. 3. Experimentally derived gas density as a function of discharge power measured at a constant pressure of 3.50 Torr.

The field maps of Fig. 2 and the gas density measurements of Fig. 3 were used to determine the average ion velocity at the cathode. Helm's (1977) precise He$^+$ ion mobilities were used for the lower currents where the E/N is less than 1500 Td. The equilibrium drift velocity $\sqrt{2eE/(m_+\pi\sigma N)}$, where m_+ is the ion mass and e the unit charge was used for higher currents. The symmetric charge exchange cross section, σ, was taken from Sinha, Lin and Bardsley (1979). These theoretical cross sections agreed with Helm's experimentally derived cross sections.

The product of the ion charge density and average ion velocity at the cathode is the ion current density J_+^0. Figure 4 is a plot of the current balance $J_+^0/(J_D-J_+^0)$ versus the discharge current density J_D. It is interesting to note that the current balance, or ratio of ion to electron current, at the cathode is ~3.3 and is essentially independent of total discharge current.

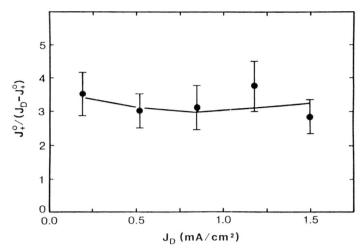

Fig. 4. Ratio of ion current to electron current at the cathode surface $J_+^o/(J_D-J_+^o)$ as a function of discharge current density. The points are the empirically derived values; the solid line is the Monte Carlo results.

2.3 Density and Temperature of the Cold Negative Glow Electrons

This section describes two laser-based diagnostics developed by Den Hartog et al. (1989) which together yield n_e^c and T_e^c for the cold trapped electrons in the negative glow. Data was taken over a range of current densities from 0.190 mA/cm^2 to 1.50 mA/cm^2, but results from the 0.846 mA/cm^2 discharge will be emphasized and analyzed in detail. Results from other current densities will be presented to illustrate trends. Two laser-based diagnostics each provide an independent determination of the product of n_e^c and a temperature dependent rate constant. The first diagnostic, based on the endothermic electron collisional transfer between low Rydberg levels, yields a relation in which n_e^c decreases with increasing T_e^c. The second diagnostic is based on the spin conversion of helium metastables,

$$He(2^1S) + e^- \rightarrow He(2^3S) + e^- + 0.79 \text{ eV} . \qquad (2)$$

This is an exothermic process and yields a relation in which n_e^c increases with T_e^c. The intersection of these two relations yields n_e^c and T_e^c for the cold electrons in the negative glow.

In the Rydberg atom diagnostic a low Rydberg level is selectively populated with the laser. The collision induced transfer rate from the level populated by the laser to a higher lying level is determined by measuring the ratio of the laser induced fluorescence (LIF) signals from the two levels. Frequency doubling of the dye laser beam is necessary for excitation of low Rydberg levels out of the 2^3S metastable level. LIF is collected in a direction perpendicular to both the discharge axis and the laser axis. A spherical lens is used to image the laser-discharge interaction region onto the 0.25 mm entrance slit of a 0.2 m monochromator which is used as a spectral filter. The ~1 nm bandpass is necessary to isolate the triplet Rydberg level fluorescence from cascade fluorescence from lower lying singlet levels. A 1P28A photomultiplier detects the filtered fluorescence. The base of the photomultiplier tube is wired for low inductance to maintain the full bandwidth of the tube. The photomultiplier signal is amplified by a factor of 100 and then digitized with a Transiac 2001S digitizer with a 4100 averaging memory interfaced to an IBM PC. Typically 5000 LIF pulses are averaged. The temporal integral of this average is the measured fluorescence F. The LIF signal from the upper Rydberg level is two to three orders of magnitude weaker than the signal from the level populated by the laser.

The Rydberg atom diagnostic works by populating a low lying Rydberg level with principle quantum number n by tuning the laser to the $2^3S \rightarrow n^3P$ transition. This is effectively populating the entire n manifold because the different ℓ levels are highly coupled by neutral collisions at 3.50 Torr (Gallagher et al. 1977). The population of each ℓ level is proportional to its degeneracy. The s level may be an exception to this since for n=5 it lies ~$3k_BT$ below the p level so that it may not be fully coupled to the rest of the manifold. Here T is the gas temperature. For the purposes of this experiment the lack of coupling of the s level is unimportant because the s population is a small fraction of the total n manifold population. The collisional transfer rate between two Rydberg manifolds n and n' (n'>n) is determined by measuring the ratio of the LIF signals $n'^3D \rightarrow 2^3P$ and $n^3D \rightarrow 2^3P$ ($F_{n'}/F_n$) when the lower lying manifold is populated with the laser.

Three transfer rates are measured. When the laser is tuned to the $2^3S \rightarrow 5^3P$ transition at 294.5 nm spatial maps of fluorescence are made from the $5^3D \rightarrow 2^3P$ (F_5) and $6^3D \rightarrow 2^3P$ (F_6) transitions at 402.6 nm and 382.0 nm respectively. F_6/F_5 yields the n=5 to n=6 collision induced transfer rate. When the laser is tuned to $2^3S \rightarrow 6^3P$ transition at 282.9 nm fluorescence is mapped from the $6^3D \rightarrow 2^3P$, $7^3D \rightarrow 2^3P$, and $8^3D \rightarrow 2^3P$ transitions at 382.0 nm, 370.5 nm, and 363.4 nm respectively. F_7/F_6 and F_8/F_6 yield the n=6 to n=7 and the n=6 to n=8 transfer rates respectively. Figure 5 shows a plot of F_8/F_6 versus distance from the cathode. The left axis of the plot is the cathode position and the right side is the anode position. The position where the linearly decreasing cathode fall field extrapolates to zero is indicated by the dashed vertical line on the plot. This is the boundary between the cathode fall (CF) and negative glow (NG).

It will be shown in the discussion that follows that the lifetimes of the Rydberg manifolds, $\tau_{n'}$, must be known in order to analyze the LIF maps. $\tau_{n'}$ is the lifetime of the manifold under discharge conditions (where the decay arises from both radiative and collisional effects). $\tau_{n'}$ is measured in situ by tuning the laser to the peak of the $2^3S \rightarrow n'^3P$ transition and collecting $n'^3D \rightarrow 2^3P$ fluorescence as described for the fluorescence mapping. The photomultiplier fluorescence signal is time-resolved and analyzed with a PAR 162/163 boxcar averager with a 2 ns gate. The results of these measurements are: $\tau_6 = 29.1$ ns, $\tau_7 = 30.8$ ns and $\tau_8 = 33.9$ ns. The uncertainty in the lifetime measurements is 10%.

The Rydberg atom diagnostic is analyzed with a straightforward rate equation approach. The laser excitation of the n manifold can be approximated as an instantaneous population excess at t=0, $N_n(0)$. The subsequent decay of the manifold population is described by a single exponential, $N_n(t) = N_n(0)\exp(-t/\tau_n)$, where τ_n is the lifetime of the n manifold in the discharge environment. Under the same conditions, the population in the n' manifold can be described by the rate equation

$$\frac{dN_{n'}(t)}{dt} = - N_{n'}(t)/\tau_{n'} + R_{nn'}N_n(t) , \qquad (3)$$

where $\tau_{n'}$ is the lifetime of the n' manifold and $R_{nn'}$ is the rate for collision induced transfer from manifold n to n', including both neutral and electron collisions. The rates due to ion collisions and due to hot electron collisions are easily shown to be insignificant (<10%). The solution to this equation is

$$N_{n'}(t) = \frac{R_{nn'}N_n(0)}{\tau_{n'}^{-1} - \tau_n^{-1}} (e^{-t/\tau_n} - e^{-t/\tau_{n'}}) . \qquad (4)$$

The ratio of the time-integrated fluorescence from the n' manifold to that from the n manifold is

$$\frac{F_{n'}}{F_n} = \frac{b_{n'}\varepsilon_{n'}\int N_{n'}(t)dt/\tau_{n'}}{b_n \varepsilon_n \int N_n(t)dt/\tau_n} \qquad (5)$$

where b_n is the branching ratio of the observed $n^3D \rightarrow 2^3P$ transition and ε is the collection efficiency of the detection system at the frequency of the $n^3D \rightarrow 2^3P$ transition. It is safe to assume that $\varepsilon_n \approx \varepsilon_{n'}$, since the frequencies of the observed transitions are nearly equal. The ratio $b_{n'}/b_n$ is given by

$$\frac{b_{n'}}{b_n} = \frac{n^2 A_{n'd-2p} \tau_{n'}}{n'^2 A_{nd-2p} \tau_n} \qquad (6)$$

where A is the radiative transition probability for the transition indicated and n^2 is the total degeneracy of the n manifold. After substitution and integrating $F_{n'}/F_n$ is given by

$$\frac{F_{n'}}{F_n} = \frac{n^2 A_{n'd-2p} R_{nn'} \tau_{n'}}{n'^2 A_{nd-2p}} \ . \qquad (7)$$

From this relation the collision induced transfer rate can be calculated from the measured ratio of fluorescence. The values for the transition probabilities are taken from Wiese et al. (1966).

Figure 5 shows the strong spatial asymmetry of the $F_{n'}/F_n$ ratio in the discharge. Throughout most of the cathode fall, to the left of the CF-NG boundary, the collisional transfer rate is low and does not vary as a function of position. The electron density is very low in the cathode

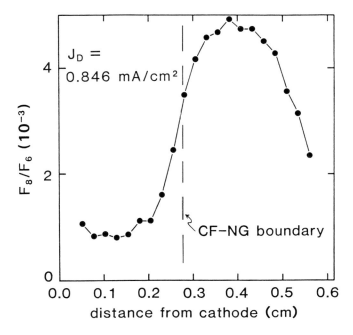

Fig. 5. Ratio of LIF signal from n=8 to that from n=6 when n=6 is excited as a function of distance from the cathode.

fall and collision induced transfers are due to neutral collisions. The neutral contribution to $R_{nn'}$ is constant as a function of position. The large enhancement of the collisional transfer rate to the right of the CF-NG boundary is due to collisions with the cold electrons trapped in the negative glow. The fact that this enhancement extends across the CF-NG boundary into the cathode fall may be due to population transfer caused by collision between hot electrons and Rydberg atoms. Table 1 lists the results of the $R_{nn'}$ determination for the three collision

Table 1. Collision induced transfer rate $R_{nn'}$ for the three collision processes studied. $(R_{nn'})_{CF}$ is interpreted as the rate due to collisions with neutrals. $(R_{nn'})_{NG} - (R_{nn'})_{CF}$ is interpreted as the rate due to collisions with the low energy electrons in the negative glow of the 0.846 mA/cm^2 discharge.

n → n'	$(R_{nn'})_{CF}$ (10^5 sec^{-1})	$(R_{nn'})_{NG} - (R_{nn'})_{CF}$ (10^5 sec^{-1})
5 → 6	2.0	3.5
6 → 7	6.0	11.1
6 → 8	1.2	4.6

processes studied. The table gives a value for $(R_{nn'})_{CF}$, which is the average over the flat portion of the data in the cathode fall, and a value for $(R_{nn'})_{NG} - (R_{nn'})_{CF}$, which is the difference between the peak in the negative glow and the cathode fall value. $(R_{nn'})_{CF}$ is the neutral collision transfer rate and $(R_{nn'})_{NG} - (R_{nn'})_{CF}$ is the rate due only to the negative glow electrons. These values result from the average of two sets of data taken several weeks apart. The two data sets reproduced within 10%.

The measured electron collision induced transfer rate can be expressed as the product of a temperature dependent rate constant $k_{nn'}(T_e^C)$, and the electron density, n_e^C. The rate constants are taken from the analytical formula given by Vriens and Smeets (1980). In evaluating this formula for manifold to manifold collision induced transfer, weighted averages are calculated for the energies of the manifolds and absorption oscillator strengths. The energies of individual levels are taken from Martin (1987). Level-to-level absorption oscillator strengths are taken from Kono and Hattori (1984) for $\ell \leq 2$, and hydrogenic values from Wiese et al. (1966) are used for higher ℓ. The hydrogenic approximation is estimated to be within a few percent of helium oscillator strengths for $\ell \geq 2$. The resulting relations, $n_e^C = k_{nn'}(T_e^C)/R_{nn'}$, are plotted as the

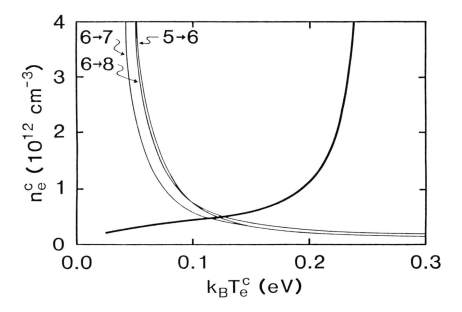

Fig. 6. Electron density versus electron temperature for the low energy electrons in the negative glow of the 0.846 mA/cm^2 discharge. The three light lines arise from the Rydberg atom diagnostic. The bold lines arise from the metastable analysis.

thin lines in Figure 6 for the three different collisional transfers studied.

The metastable atom diagnostic is carried out under the same discharge conditions as the Rydberg atom diagnostic. This diagnostic is also the study of a collision induced transfer, but in this case the reaction is the exothermic metastable spin conversion reaction of Equation 2. The transfer rate is determined by mapping the steady state densities of helium 2^1S and 2^3S metastables, and then calculating a fit to the density maps based on an analysis of the metastable transport and kinetics.

The metastable densities are measured by first mapping the relative densities using LIF, and then an absorption measurement is made to put an absolute scale on the maps. The 2^1S metastables are mapped by driving the $2^1S \rightarrow 3^1P$ transition at 501.6 nm with the laser and observing $3^1D \rightarrow 2^1P$ fluorescence at 667.8 nm. The 2^3S metastables are mapped by driving the $2^3S \rightarrow 3^3P$ transition at 388.9 nm with the laser and observing $3^3S \rightarrow 2^3P$ fluorescence at 706.5 nm. Populations of 3p and 3d levels are coupled by neutral collisions (Wellenstein and Robertson 1972, Dubreuil and Catherinot 1986). Fluorescence collection and signal processing is the same as for the Rydberg atom diagnostic except for the substitution of interference filters for the monochromator as a spectral filter.

An absorption measurement is made at the peak of the spatial profile to put an absolute scale on the density maps. The laser is scanned through the same transition used for the LIF while the transmitted laser power is monitored with a photodiode detector. The signal is processed with a boxcar averager and plotted on a stripchart recorder. To simplify the analysis of the absorption lineshape the laser power is nonsaturating and the bandwidth is reduced to 500 MHz by introduction of an extra-cavity plane parallel etalon. Metastable densities are determined from the integral over frequency of the natural log of the transmittance. Ten absorption measurements are averaged to determine each density. The results of the metastable density measurements are shown as points in Figure 7.

The net rate for the metastable spin conversion reaction is determined from an analysis of the density maps. The transport and kinetics of 2^1S and 2^3S metastables in the discharge are modeled using a pair of balance equations,

$$D_s \frac{\partial^2 M_s}{\partial z^2} - \beta M_s^2 - \beta M_s M_t - \gamma M_s N - \kappa n_e^c M_s + P_s = 0 \qquad (8)$$

$$D_t \frac{\partial^2 M_t}{\partial z^2} - \beta M_t^2 - \beta M_s M_t + \kappa n_e^c M_s + P_t = 0 \qquad (9)$$

where M is the metastable density, N is the density of the ground state atoms, D is the diffusion coefficient, P is the production rate per unit volume, γ is the rate constant for singlet metastable destruction due to collisions with ground state atoms, β is the rate constant for destruction of metastables due to metastable-metastable collisions, κ is the rate constant for net destruction of singlet metastables due to low energy electron collisions, and s and t subscripts indicate singlet and triplet metastables respectively. The first three loss terms in each equation are loss terms arising from diffusion and collisions between metastables. There is an additional loss term in the singlet equation arising from collisions with ground state atoms. The fifth loss term in the singlet equation, and a corresponding gain term in the triplet equation, is due to the spin conversion of metastables. Values of D, N, γ and β are taken from Phelps (1955). The temperature dependence of D was taken from Buckingham and Dalgarno (1952) and that of γ from Allison, Brown and Dalgarno (1966). In the original reporting of this diagnostic, there was a small error in the temperature dependence of D. The error was ~15% for this discharge, and has been corrected here. The correction did not affect the values of κn_e^c reported in Den Hartog et al. (1988) but did improve agreement between metastable production deduced from the experiment and from Monte Carlo simulations (Fig. 13 of Den Hartog et al. 1988).

The coupled equations are solved numerically for $M_s(z)$ and $M_t(z)$ and compared with the experimental results. For the purposes of this model, the spatial dependence of the production terms is assumed to be a fundamental diffusion mode and that of n_e^c a step function: zero in the cathode fall and constant in the negative glow. The assumption about the spatial dependence of the production term is justified by Monte Carlo simulations described in the next section. Diffusion modes up to tenth order are included in the calculation to describe the spatial asymmetry of the 2^1S metastables. The values of P_s, P_t and κn_e^c are varied until a good fit to the experimental metastable maps is obtained. The dashed curves in Figure 7 are the calculated metastable maps. By the method outlined above, κn_e^c is found to be 7×10^4 sec^{-1}.

The rate constant for metastable spin conversion, κ_m, was measured by Phelps (1955) for room temperature electrons. The temperature dependence of the rate constant is $(T_e^c)^{-1/2}$ because the cross section scales as the inverse of the electron energy (Fon et al. 1981). The value of κn_e^c determined from the metastable analysis must be considered an effective rate for metastable spin conversion since it accounts for both the forward and reverse contributions of the reaction (Eq. 2). Setting κn_e^c equal to the expression for the forward rate minus the reverse rate we obtain

$$\kappa n_e^c = n_e^c \left[\kappa_m \sqrt{\frac{0.025 \text{ eV}}{k_B T_e^c}} - \left(\frac{M_t}{3M_s}\right) \kappa_m \sqrt{\frac{0.025 \text{ eV}}{k_B T_e^c}} \exp\left(-0.79 \text{ eV}/k_B T_e^c\right) \right]. \quad (10)$$

This expression yields a second relation between n_e^c and T_e^c which is plotted as the bold line in Figure 6. A natural upper bound on T_e^c arises from the above analysis due to the equilibration of the reaction. The singlets will only be depopulated under the conditions that the forward rate is greater than or nearly equal to the reverse rate.

The two diagnostics discussed above each yield independent relations between n_e^c and T_e^c for the cold electrons in the negative glow. The intersection of these relations, plotted in Figure 6, yields $n_e^c = 5 \times 10^{11}$ cm^{-3} and $k_B T_e^c = 0.12$ eV for the electrons in the negative glow of a 261 V, 0.846 mA/cm^2 discharge in 3.50 Torr helium. The uncertainty in the Rydberg atom diagnostic is primarily due to the uncertainty in the rate coefficient at electron temperatures near the reaction threshold. We estimate this uncertainty to be − 50%, +100%. This conservative estimate is based on a comparison by Vriens and Smeets (1980) of their rate coefficients with experimental measurements for n=13 to n=14 and n=15 collision induced transfer rates for T_e near threshold. The uncertainty in the metastable atom diagnostic is due mainly to the assumption of a

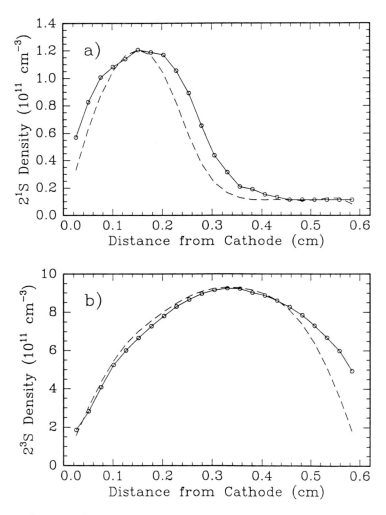

Fig. 7. 2^1S and 2^3S metastable densities as a function of distance from the cathode in the 0.846 mA/cm^2 discharge. The points are experimental measurements. The dashed curves are calculated densities using the fitting process described in the text.

simple two level system. The collisional and radiative coupling of the 2^1S with the 2^1P is not entirely negligible. Particularly the $2^1P \rightarrow 2^1S$ radiation will be trapped in the cathode fall but not in the negative glow. The result will be a spatially asymmetric feeding of the 2^1S population in the negative glow. The effect will be to push the metastable analysis curve to higher n_e^c by as much as a factor of three

(Den Hartog 1989). With these uncertainties in mind, the ranges of electron density and temperature consistent with the results of the two diagnostics are $2.5 \times 10^{11} < n_e^c < 8 \times 10^{11}$ and $0.08 \text{ eV} < k_B T_e^c < 0.2 \text{ eV}$.

Figure 8 shows the measured values of n_e^c and $k_B T_e^c$ as a function of current density for the 0.846 mA/cm^2 and other discharges studied. It is very important to note that $k_B T_e^c$ depends linearly on current density and that values of $k_B T_e^c$ in the 0.1 eV to 0.2 eV range are typical.

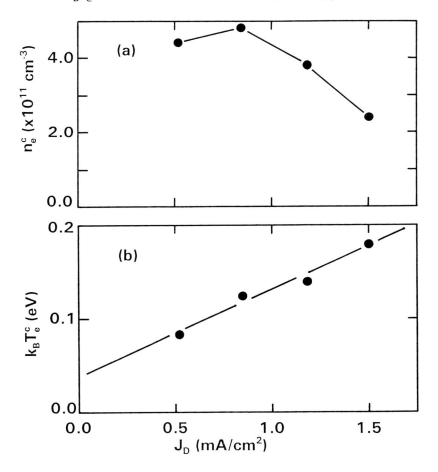

Fig. 8. (a) Density and (b) temperature of the cold negative glow electrons as a function of discharge current density. The data points are simply connected in (a), and are least square fitted to a straight line in (b).

3. MONTE CARLO SIMULATIONS

Monte Carlo simulations are now widely used to study the behavior of charged particles in weakly ionized gases. These simulations are not as elegant as a numerical solution of the Boltzmann equation, and they do require considerably more computer time than numerical solutions. The computer time issue is steadily decreasing in significance; all of the simulations described in this work were executed on a personal computer. Monte Carlo simulations are very easy to code, and they can be made quite realistic if sufficient high quality data is available on cross sections including angular distributions for elastic scattering, for excitation, and for ionization. It is straightforward to include almost any amount of detail in a simulation which is justified by the data base of cross sections. Any well defined physical quantity can be determined in a Monte Carlo simulation.

Monte Carlo simulations are used in this work to study electron avalanches in the cathode fall and negative glow regions. There is no need for Monte Carlo simulations of ion transport in the cathode fall region because we have the exact analytic solutions to the Boltzmann equation for ions (Lawler 1985). The Monte Carlo simulations include the empirical field distributions and gas densities for the cathode fall region as given in Section 2. A uniform (nonreversed) field of 1.0 V/cm or 10.0 V/cm is assumed for the negative glow region. There is actually a reversal of the field direction near the boundary between the cathode fall and negative glow regions. The reversal produces a potential energy well which holds the cold electron gas of the negative glow. The reversal also results in a back diffusion (or drift) toward the anode of ions produced in the negative glow. Unfortunately it is difficult to include a field reversal in the Monte Carlo simulations. It is also difficult to study the cold trapped electrons in the negative glow using a Monte Carlo simulation. However, the actual magnitude and direction of the weak field in the negative glow region has very little effect on excitation and ionization in the negative glow which is produced by high energy or hot electrons injected from the cathode fall region.

The Monte Carlo simulations are three dimensional in the sense that they include angular scattering, but they do not include edge effects due to fringing of the discharge. One-dimensional infinite plane parallel geometry is assumed. The most convenient coordinates are T, μ, and z where T is the electron's kinetic energy, μ is the cosine of the angle between the electron's velocity and acceleration, and z is the distance from the cathode. The acceleration $\bar{a}$ of magnitude $|eE/m_e|$ where e is the unit charge, E the electric field, and m_e the electron mass, is in the positive z direction. The quantity s is a distance along an electron's trajectory between collisions, and W is the electron's potential energy.

Two quantities are conserved while the electron is in free flight between collisions, the transverse kinetic energy T* where

$$T^* = T(1-\mu^2) \qquad (11)$$

and the total energy T + W(z). An electron is launched after a collision with initial conditions T_1, μ_1, and z_1. Once a single chosen coordinate is determined immediately before the next collision, such as T_c, μ_c, or z_c, the other two coordinates are determined from conservation equations. The change in the chosen coordinate is determined from a pseudo random number in a fashion that reflects the electron's scattering probability along its trajectory. By tracking large numbers of electrons one can determine the same information embodied in a solution to the Boltzmann equation: a phase distribution function and all moments of it. The enormous advantage of a null collision Monte Carlo technique is that it makes it possible to relate the change in the chosen coordinate between collisions to the pseudo random number without performing a numerical integration of scattering probability along the trajectory. Numerical integration is a slow process and is to be avoided if possible.

The null collision technique used in this investigation was developed by Boeuf and Marode (1982) for nonuniform fields. The probability density function $q(s_c)$ is used to describe the differential probability of a collision occuring in a interval ds about s_c,

$$dp = q(s_c)ds = \exp\{-\int_0^{s_c} N\sigma[T(s)]ds\} \, N\sigma[T(s_c)]ds. \qquad (12)$$

The electron starts with s=0 after the last collision at coordinates T_1, μ_1, z_1. The cross section σ includes elastic, excitation, and ionization scattering. It is obviously very dependent on the electron kinetic energy T. The exponential in the definition of q is the probability of the electron surviving without a collision to s_c, and the remaining factor is the probability of scattering at s_c. The cumulative distribution function

$$p(s_c) = \int_0^{s_c} dp = \int_0^{s_c} q(s)ds = 1-\exp\{-\int_0^{s_c} N\sigma[T(s)]ds\} \qquad (13)$$

is a probability of a collision occuring before the electron reaches s_c. This function should be set equal to a pseudo random number r_1 on the interval from 0 to 1.0 and solved for s_c to determine T_c or μ_c or z_c in the Monte Carlo simulation,

$$\int_0^{s_c} N\sigma[T(s)]ds = -\ln(1-r_1) = R. \qquad (14)$$

Following Boeuf and Marode (1982) we note

$$ds = dz/\mu \qquad (15)$$

and

$$m_e a \, dz = dT \tag{16}$$

thus

$$\int_{T_1}^{T_c} \frac{N\sigma(T)}{m_e a \, \mu} dT = \int_{T_1}^{T_c} \frac{N\sigma(T)\sqrt{T}}{m_e a(T-T^*)^{1/2}} \, \text{sgn}(\mu) dT = R. \tag{17}$$

Note that the acceleration a, which is a function of position, must be expressed as a function of T in order to evaluate this integral. Of course the null collision technique avoids this numerical integration. There is an upper bound, B_i, of $N\sigma(T)\sqrt{T}/(am_e)$ in the i'th spatial bin defined by the interval z_i to z_{i+1}. The integration of Eq. 17 is straightforward if a null scattering process is included, with a cross section chosen such that

$$N[\sigma(T) + \sigma_{\text{null}}(T)] \sqrt{T}/(am_e) = B_i \tag{18}$$

throughout the i'th bin from z_i to z_{i+1}. Null collisions, when they occur, do not change the electron's velocity. This algorithm cannot deal with a field reversal because $N\sigma(T)\sqrt{T}/(am_e)$ diverges at a field reversal. The remaining complications in the integral of Eq. 17 are due to $\text{sgn}(\mu)$ and to the steps in the integrand caused by $B_{i-1} \neq B_{i+1}$.... These complications are handled by appropriate branches and loops in the program. First we consider a non-negative μ_1. The equation

$$2B_i(\sqrt{T-T^*} - \sqrt{T_1-T^*}) = R \tag{19}$$

is solved, and the solution T is compared to T_{i+1}^b, the kinetic energy at the boundary between the i'th and i'th+1 bin at z_{i+1}. If T is less than T_{i+1}^b, then $T=T_c$ and a collision occurs in the i'th bin. Otherwise the equation

$$2B_{i+1}(\sqrt{T-T^*} - \sqrt{T_{i+1}^b - T^*}) = R - 2B_i(\sqrt{T_{i+1}^b - T^*} - \sqrt{T_1-T^*}) \tag{20}$$

is solved, and the solution T is compared to T_{i+2}^b. This procedure is iterated until the bin and the kinetic energy T_c immediately before the collision is determined. A similar procedure is followed for negative μ_1 with the additional possibility of the electron reversing direction before colliding. For example we consider an electron with negative μ, but without sufficient longitudinal kinetic energy T_1-T^* to reach z_i. The equation

$$-2B_i(\sqrt{T-T^*} - \sqrt{T_1-T^*}) = R \tag{21}$$

can be solved for T only if R is sufficiently small. A large value of R implies that the electron passes through the turning point where $T=T^*$ and $\mu=0$; thus the equation

$$2B_i(\sqrt{T-T^*} + \sqrt{T_1-T^*}) = R \tag{22}$$

is solved for T.

The type of collision is determined by choosing a second pseudo random number r_2 in the interval from 0 to 1.0. The 0 to 1.0 interval is divided into segments, each of length equal to the probability of a particular type of collision at energy T_c. The cross section σ is a sum of the elastic cross section σ_{el}, the ionization cross section σ_{ion}, and various excitation cross sections. The first segment is of length $\sigma_{null}(T_c)/[\sigma_{null}(T_c) + \sigma(T_c)]$, the second is of length $\sigma_{el}(T_c)/[\sigma_{null}(T_c) + \sigma(T_c)]$, the third is of length $\sigma_{ion}(T_c)/[\sigma_{null}(T_c) + \sigma(T_c)]$, and so on for each excitation cross section. The segment containing r_2 determines the type of collision.

The electron's coordinates T_2 and μ_2 immediately after the collision must be determined. The procedure is dependent on the type of collision. Null collisions, of course, result in $T_2 = T_c$ and $\mu_2 = \mu_c$. The determination of T_2 and μ_2 for real collisions requires differential cross sections. The sources of the cross sections used in the simulations are given during the discussion of T_2 and μ_2.

The total and differential elastic scattering cross section for electron impact on He calculated by LaBahn and Callaway (1969-70) are used in the simulations. The coordinate μ_2 after an elastic collision is calculated from the scattering angle χ and the azimuthal scattering angle ϕ using the equation

$$\mu_2 = \mu_c \cos\chi + \sqrt{1-\mu_c^2} \sin\chi \cos\phi. \tag{23}$$

The scattering angles χ and ϕ are determined from two more pseudo random numbers using the equations

$$r_3 = \frac{2\pi}{\sigma_{el}} \int_0^\chi \frac{d\sigma_{el}}{d\Omega} \sin\Theta \, d\Theta \tag{24}$$

and

$$r_4 = \phi/2\pi. \tag{25}$$

Equation 24 can be solved for χ by numerical integration of LaBahn and Callaway's (1969-70) differential cross section. Approximations such as

$$\frac{d\sigma_{el}}{d\Omega} = A[1+g(T_c)\sin^2(\chi/2)]^{-1} \tag{26}$$

are particularly convenient because χ can be evaluated without numerical integration. The function $g(T_c)$ is roughly quadratic in T_c. The kinetic energy after the collision is

$$T_2 = T_c [1-(2m_e/m_o)(1-\cos\chi)], \tag{27}$$

where m_o is the atom mass.

The analytic expressions given by Alkhazov (1970) for excitation cross sections are used in the simulations. Excitation of the n=2 through n=5 singlet and triplet s and p levels is included in the simulations. Excitation of the n=3 through n=5 singlet and triplet d levels is also included. There is very little experimental or theoretical information on differential excitation cross sections for electrons on He. Thus we use an isotropic scattering approximation for excitation,

$$\mu_2 = 2r_3 - 1.0. \tag{28}$$

The kinetic energy immediately after an excitation collision is

$$T_2 = T_c - W_{exc} \tag{29}$$

where W_{exc} is the appropriate excitation energy. Recoil is neglected in this approximation.

A fraction of the atoms excited to the 3p levels and above are associatively ionized. An accurate determination of the fraction requires accurate associative ionization cross sections, cross sections for excitation transfer due to atom and electron collisions, and radiative decay rates. On the basis of what information is available, we estimate that 25% of the atoms excited to the 3p levels and above are associatively ionized at 3.5 Torr (Wiese et al. 1966, Wellenstein and Robertson 1972, Dubreuil and Catherinot 1986). The electrons released in associative ionization contribute to the avalanche; they are assumed to have an initial kinetic energy of 1.0 eV and an isotropic angular distribution. The remaining 75% of the atoms excited to the 3p and higher triplet levels, and all of the atoms excited to the 3^3S and 2^3P levels are assumed to radiate to the 2^3S metastable level. This cascade contribution is combined with direct excitation of the 2^3S level to determine the total production rate for 2^3S metastables. The total production rate for 2^1S metastables includes direct excitation and a cascade contribution of 19% of the excitation to the 3p and higher singlet levels. The total production rate for the 2^1P resonant level includes direct excitation and a cascade contribution of all of the 3^1S excitation and 56% of the excitation to the 3p and higher singlet levels. These assumptions are based on a significant ℓ mixing of 3p and higher level populations due to collision with ground state atoms.

In a vacuum nearly all of the atoms excited to the 2^1P resonant level decay to the ground state via VUV emission. Radiation trapping reduces the effective decay rate to 1.3×10^6 sec^{-1}, which is comparable to the vacuum decay rate of 1.98×10^6 sec^{-1} for the $2^1P \rightarrow 2^1S$ infrared branch at 2μm (Wiese et al. 1966). However the $2^1P \rightarrow 2^1S$ branch is also trapped in the cathode fall region of our discharge experiments, thus we do not include any of the 2^1P excitation in calculating a total production rate for 2^1S metastables. This approximation is not completely satisfactory because the $2^1P \to 2^1S$ branch is not highly trapped in the negative glow where the 2^1S density is suppressed by low energy electron collision. The 2^1P level is feeding population to the 2^1S level through a collisional-radiative coupling, and the 2^1S level is feeding population to the 2^3S through a collisional coupling. The net result is that the Monte Carlo simulation produces a metastable production rate which is somewhat lower than the empirical rate.

The analytic expression given by Alkazov (1970) for the total ionization cross section is used in the simulations. Alkhazov (1970) also provides an analytic expression for the differential cross section with respect to energy. The kinetic energy of the outgoing scattered electrons is determined from a pseudo random number using the expression,

$$r_3 = \frac{1}{\sigma_{ion}} \int_0^{T_2} \frac{d\sigma_{ion}}{dT} dT. \tag{30}$$

The outgoing ejected electron has kinetic energy

$$T_2' = T_c - T_2 - W_{ion} \tag{31}$$

where W_{ion} is the ionization potential. The coordinates μ_2 and μ_2' are determined using the same assumptions used by Boeuf and Marode (1982): (1) the incident, scattered, and ejected electron velocities are coplanar, and (2) the scattered and ejected electron velocities are perpendicular. These assumptions result in

$$\cos\chi = \sqrt{T_2/(T_c - W_{ion})} \tag{32}$$

and

$$\cos\chi' = \sqrt{T_2'/(T_c - W_{ion})}. \tag{33}$$

The azimuthal scattering angle ϕ is determined by a pseudo random number as in Eq. 25. Equation 23 is used to determine μ_2. The coordinate μ_2' is calculated using

$$\mu_2' = \mu_c \cos\chi' - \sqrt{1-\mu_c^2} \sin\chi' \cos\phi. \tag{34}$$

The justification for this angular distribution is strong only at very high energy, but it does represent a reasonable attempt to introduce anisotropic scattering in ionizing collisions.

The details of the electron interactions with the cathode and anode must be specified. The kinetic energy of the electrons emitted from the cathode is assumed to be 5.0 eV and they are assumed to have a random angular distribution in the forward direction. The use of a more realistic energy distribution for electrons emitted at the cathode does not affect ionization or excitation in the Monte Carlo simulations. Some electrons are reflected back to the cathode by elastic scattering from He atoms before they suffer an inelastic collision. Electrons do not have sufficient total energy to reach the cathode after an inelastic collision. The percentage of the gross electron emission which is back scattered to the cathhode varies from 19% at the lowest current density (a near normal cathode fall) to 3.6% at the highest current density (a highly abnormal cathode fall). All Monte Carlo results are presented in terms of net electron emission from the cathode. The Monte Carlo simulation assumes no reflection from the anode, each electron is absorbed immediately when it reaches the anode. This approximation will cause the Monte Carlo simulation to predict excitation and ionization rates in the negative glow which are lower than in experiment. The discordance is significant only in the case of the highly abnormal glow discharge at 1.50 mA/cm^2. An avalanche ends when all of the electrons in the avalanche reach the anode. Trapped electrons cannot be included in the Monte Carlo simulation because the simulation will not conclude.

Figure 9 includes Monte Carlo histograms giving the number of ionization and excitation events per (net) electron emitted from the cathode as a function of distance from the cathode. These histograms are for field distributions and gas densities corresponding to 0.846 mA/cm^2. The spatial dependence of the ionization and excitation rates are all described by a roughly symmetric function which peaks near the cathode fall-negative glow boundary. The simulations show that the hot "beam" electrons in He penetrate beyond the cathode fall-negative glow boundary by a distance comparable to d_c, the thickness of the high field cathode fall region. The penetration distance will scale as the inverse of the gas density in a fashion similar to d_c. Although the hot electrons are often described as beams, their degree of anisotropy depends strongly on the cathode fall voltage and on the energy of the electrons.

Figure 10 is the hot electron density as a function of distance from the cathode. As expected, the density of hot electrons is rather insensitive to the negative glow field. This field is expected to be in the ~ 1 V/cm range or less. An order of magnitude increase in the negative glow field lowered the negative glow hot electron density by ~30%. The absolute scale on the density is based on the empirical current balance of 3.13 at the cathode of this 0.846 mA/cm^2 discharge. The current balance, which is the ratio of ion to electron current, has an uncertainty of 20%.

Figure 11 is the average kinetic energy, $\langle m_e v^2/2 \rangle_e$, of the hot electrons as a function of distance from the cathode. These results are independent of the current balance at the cathode, and only weakly

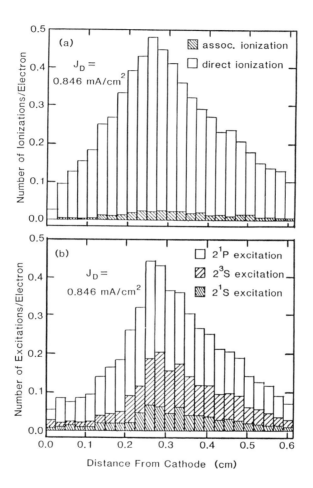

Fig. 9. Monte Carlo histograms showing the number of (a) ionization events and (b) excitation events as a function of distance from the cathode. The histogram for total ionization is subdivided into associative and direct ionization. The histogram for total excitation is divided into 2^1S, 2^3S, and 2^1P excitation. The simulation used an empirical cathode fall field of 1870 V/cm at the cathode, decreasing linearly with distance from the cathode, and extrapolating to zero at a distance of 0.282 cm from the cathode. A uniform (nonreversed) field of 1.0 V/cm is used in the negative glow.

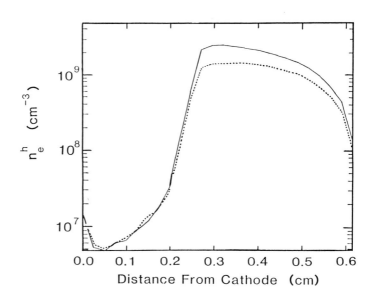

Fig. 10. Density of the hot electrons as a function of distance from the cathode determined in Monte Carlo simulations. The solid curve is from a simulation with a 1.0 V/cm field in the negative glow. The dotted curve is for a 10.0 V/cm field in the negative glow. Both simulations used the same empirical cathode fall field for the 0.846 mA/cm^2 discharge.

dependent on the assumed field in the negative glow. The energy loss rates by hot electrons due to elastic collisions with neutrals and due to Coulomb collisions with cold electrons are comparable, but both are unimportant in the power balance of the hot electrons. The average energy of the hot electrons increases only 6% when energy loss due to elastic collisions is removed from the Monte Carlo code. (We show in Section 5 that energy transfer from the hot electrons to the cold electrons is important in the power balance of the cold electrons.)

Figure 12 is a histogram in the negative glow (0.418 cm from the cathode) of the speed distribution, $\int f_e 2\pi v^2 d\mu$, where f_e is the full phase space electron distribution, v is the speed, and μ the cosine of the angle between the electron velocity and discharge axis. This histogram is from a simulation with a 1.0 V/cm field in the negative glow. A Maxwellian

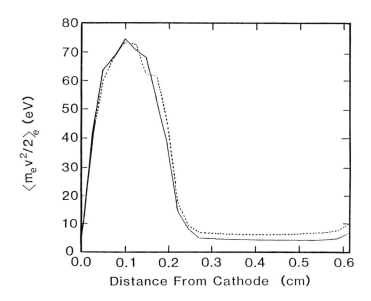

Fig. 11. Average kinetic energy of the hot electrons as a function of distance from the cathode. Note that an order of magnitude increase in the negative flow field from 1.0V/cm (solid curve) to 10.0V/cm (dotted curve) only increased the average energy in the negative glow by ~33%.

speed distribution normalized to the same density and for a corresponding electron temperature ($k_B T_e^h = (2/3)\langle m_e v^2/2 \rangle_e = 3.05$ eV) is included on the plot. The Monte Carlo distribution peaks at lower speed and has an excess of high speed electrons in comparison to the Maxwellian. The energy transfer rate due to Coulomb collisions has a rather weak energy dependence which justifies an approximation in Section 5. These hot electrons are approximated as Maxwellian in a calculation of energy transfer to the cold electrons from Coulomb collisions.

The absolute accuracy of the Monte Carlo simulations must be estimated in order to compare the simulations to experiment. Statistical uncertainty is not large because several thousand complete avalances were simulated. Alkhazov's (1970) analytic expressions for the He cross sections are based on theoretical and experimental work by many groups. He estimates that the uncertainty is ±25% at low energy, and ±5% at high energy where

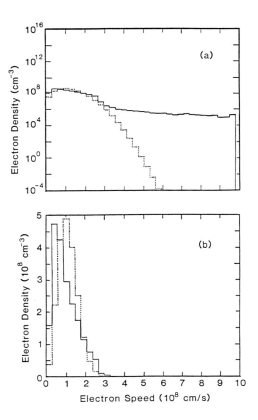

Fig. 12. Histogram of the speed distribution of the hot electrons in the negative glow. The solid line is from the Monte Carlo simulation with a 1.0V/cm field in the negative glow. The dotted line is a Maxwellian speed distribution for the same density and average energy. The cutoff of the Monte Carlo histogram corresponds to the maximum electron energy which is determined by the discharge voltage. a) Semilogarithmic plot. b) Linear plot.

a Born approximation is reliable. We believe his uncertainties are reasonable and we estimate that the Monte Carlo results have an uncertainty of $\pm 25\%$.

4. IONIZATION BALANCE IN THE NEGATIVE GLOW

The Monte Carlo simulations described in Section 3 can be compared to the empirical current balance from Section 2. All of the ions produced in the cathode fall region are driven to the cathode by the large electric field. The fate of ions produced in the negative glow is more difficult to predict a priori, but is explored in this section.

The cumulative number of ionization events per electron emitted from the cathode is determined by summing events in each bin of the histogram, Fig. 9, from the surface of the cathode to the cathode fall-negative glow boundary where the field extrapolates to zero. This sum is directly comparable to the empirical current balance. It could be less than the current balance if ions carry a significant fraction of the discharge current across the cathode fall-negative glow boundary, but it should not be greater than the empirical current balance. As indicated in Fig. 4, we find good agreement between the empirical current balance and the Monte Carlo simulations. This observation leads to the conclusion that ionization in the negative glow must be balanced either by electron-ion recombination in the gas phase or by losses to the anode. The experiments described in Section 2 show that the temperature of the cold electrons, T_e^c, is significantly above gas temperature. This indicates, as will be discussed in Section 6, that electron-ion recombination in the gas phase is probably not important. We suggest that there is a field reversal near the cathode fall-negative glow boundary and that ionization is balanced by a process analogous to ambipolar diffusion. This "ambipolar-like" diffusion causes a drift of ions to the anode with an electron current exceeding the ion current by the discharge current.

Den Hartog et al. (1988) used the empirical current balance from Section 2 and the Monte Carlo results of Section 3 to determine the spatially averaged ionization rate per unit volume. This can be combined with an estimate of the loss rate per unit volume due to ambipolar like diffusion. The resulting balance equation provides an additional constraint on n_e^c and T_e^c in the negative glow. We argue that the temperature of the cold electrons, which balance the ion space charge density in the negative glow, must to a substantial extent, control the ambipolar-like diffusion. The resulting balance equation is

$$\int_{d_c}^{d_c+d_{ng}} P_i dz/(d_{ng}) = \mu_+ k_B T_e^c (\pi/2d_{ng})^2/e \qquad (35)$$

where P_i is the ionization rate per unit volume, d_{ng} is the thickness of the negative glow, and where μ_+ is the low field ion mobility. This expression is plotted along with results from the metastable experiment in Fig. 13. The intersection of the two curves is at a somewhat higher value of $k_B T_e^c$ and n_e^c than that from Section 2.3. However, if one keeps in mind that the results of Section 2.3 have significant uncertainty, and that the results here also have uncertainty, then the factor of two agreement is satisfactory. Part of the discordance is undoubtedly due to

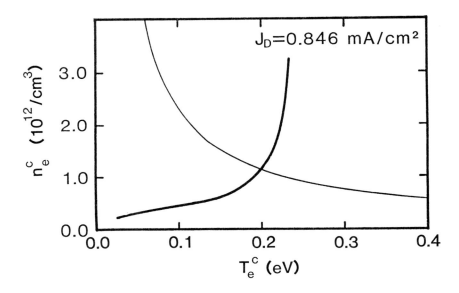

Fig. 13. Cold electron density vs electron temperature in the negative glow. The thick line is the relationship from the analysis of the metastable maps. The thin solid line is from the ionization balance equation based on "ambipolar-like" diffusion.

the use of the simplest approximation for ambipolar diffusion in describing the transport of ions in the negative glow. The nonhydrodynamic electron distribution function with a significant excess of high energy electrons in the negative glow affects the accuracy of this approximation.

5. POWER BALANCE OF THE COLD NEGATIVE GLOW ELECTRONS

In this section different terms in the power balance of the cold negative glow electrons are explored. Power loss or the cooling rate is calculated for two processes: elastic collisions with neutral atoms and Coulomb collisions with ions. Power gain or the heating rate is also calculated for two processes: Coulomb collisions with hot electrons, and superelastic collisions with metastable atoms. All cooling and heating rates will be calculated per cold electron and expressed in units of eV/sec. Where necessary experimental results such as the density ($n_e^c = 5 \times 10^{11} \text{cm}^{-3}$) and the temperature ($k_B T_e^c = 0.12$ eV) of the cold negative glow electrons are used. Monte Carlo simulations are used to determine the density and average energy of the hot negative glow electrons.

The fraction of the electron kinetic energy transferred due to recoil in an elastic collision with a neutral atom at rest is $2(1-\cos\chi)m_e/m_o$ where m_e is the electron mass, m_o is the atom mass, and χ is the scattering angle. Thus the elastic energy transfer cross section is proportional to the momentum transfer cross section, σ_{mt}, which is almost independent of energy for low energy electrons on He atoms. The momentum transfer cross section is $6.6 \times 10^{-16} \text{cm}^2$ for 1 eV and $6.5 \times 10^{-16} \text{cm}^2$ for 4 eV electrons on He (LaBahn and Callaway 1969-70). Precision swarm experiments by Crompton et al.(1970), indicate that σ_{mt} decreases slightly for electron energies below 1 eV; a value of $6 \times 10^{-16} \text{cm}^2$ is used. Assuming a Maxwellian electron energy distribution and neglecting the thermal motion of the atoms, the cooling rate per electron is $8\sigma_{mt} N (k_B T_e^c)^{3/2} \sqrt{2m_e}/(m_o\sqrt{\pi})$ where N is $1.03 \times 10^{17} \text{cm}^{-3}$, the neutral atom density. This expression yields a cooling rate of 9.4×10^4 eV/sec.

If the effect of a non-zero gas temperature ($k_B T = 0.028$ eV) is included, then the cooling rate is $8\sigma_{mt} N (k_B T_e^c - k_B T)(2k_B T_e^c/m_e + 2k_B T/m_o)^{1/2} m_e m_o/[(m_e+m_o)^2 \sqrt{\pi}]$. The derivation of this formula is straightforward but somewhat tedious. It yields a cooling rate per electron of 7.2×10^4 eV/sec.

The cooling rate for Coulomb collisions of electrons with ions, $4\sqrt{2\pi}\, n_e^c e^4 (k_B T_e^c - k_B T_i)(m_o k_B T_e^c + m_e k_B T_i)^{-3/2} \sqrt{m_e m_o} \ln(2/\chi_m)$, is derived by Longmire (1963). Here $\ln(2/\chi_m)$ is taken to be 10, the ion temperature T_i is the same as the gas temperature, and e is the unit charge. The ion mass is negligibly different from the atom mass m_o. A cold electron in the negative glow loses 1.3×10^4 eV/sec in Coulomb collisions with ions.

The heating rate due to Coulomb collisions between hot electrons and cold electrons is $4\sqrt{2\pi}\, n_e^h e^4 (k_B T_e^h - k_B T_e^c)(k_B T_e^h + k_B T_e^c)^{-3/2} \ln(2/\chi_m)/\sqrt{m_e}$ as derived by Longmire (1963). The hot electron density ($n_e^h = 1.56 \times 10^9 \text{cm}^{-3}$) and temperature ($k_B T_e^h = 3.05$ eV) are spatial averages over the negative glow of the Monte Carlo results from Sec. 3. The hot electrons, which are somewhat non-Maxwellian, are here approximated as Maxwellian with $k_B T_e^h$ equal to 2/3 of their average kinetic energy. The heating rate per cold electron is 7.1×10^4 eV/sec.

Superelastic collisions of cold electrons with metastable atoms provide some heating of the cold electrons. Collisions in which metastable atoms are de-excited to the ground state, are not important. The rate constants of a few $\times 10^{-9} \text{cm}^3/\text{sec}$ to $10^{-8} \text{cm}^3/\text{sec}$ are small in comparison with other rate constants (Nesbet 1980). Furthermore, the outgoing electron has ~20 eV of kinetic energy. This energetic electron will escape from the negative glow before it shares a significant fraction of its kinetic energy with the cold electrons through Coulomb collisions. The metastable spin conversion reaction of Eq. (2) is a more effective, but not dominant, heating mechanism for the cold electrons. For electrons at ambient temperature ($k_B T_e$ = 0.026 eV), the rate constant of this reaction is 3.2×10^{-7} cm^3/sec (Phelps 1955). This reaction does severely suppress the singlet metastable density in the negative glow. The empirical singlet metastable density M_s, and effective reaction rate Kn_e^c which includes the reverse reaction, are used to estimate the heating rate per cold electron 0.79 eV M_s Kn_e^c/n_e^c. This expression yields 0.13×10^4 eV/sec.

The power balance of the cold negative glow electrons is dominated by cooling due to energy lost in elastic collisions with neutral atoms, and by heating due to energy gain in Coulomb collisions with hot electrons in the negative glow. The hot electron density increases linearly with current density. This explains why $k_B T_e^c$ increases linearly with current density and extrapolates to $k_B T$ (gas temperature) at low current densities as shown in Fig. 8.

6. LIGHT FROM THE NEGATIVE GLOW

The negative glow region is typically an order of magnitude or more brighter than the cathode fall region (Crookes dark space) in a cold cathode glow discharge. The combination of: (1) a detailed distribution function for hot negative glow electrons, (2) the spatial maps of the absolute metastable densities from the experiment (Den Hartog et al. 1988), and (3) recently measured electron impact cross sections for excitation from the 2^3S metastable level to higher levels (Rall et al. 1989), is used in this section to assess the role of multistep excitation in producing light from the negative glow. Multistep excitation is important in the negative glow. Electron-ion recombination is not required to explain the brightness of the negative glow.

One of the strongest visible lines in the HeI spectrum is the line at 587.6 nm from the 3^3D level. This level decays only by the 587.6 nm transition to the 2^3P level. In this discussion excitation transfer to and from the 3^3D level from and to other nearby levels will be neglected. The calculated single-step, and single-step plus two-step excitation rates per unit volume are shown in Fig. 14. The only important two-step excitation process is from the 2^3S metastable level, because the density of 2^1S metastables in the negative glow is far lower.

The single-step excitation rate per unit volume is only slightly asymmetric about the cathode fall-negative glow boundary. This slight asymmetry is due to the sharp peak in the energy dependence of the cross section just above threshold. The energy dependence is typical of cross sections for spin changing collisions, such as this 3^3D excitation from the 1^1S ground level. Monte Carlo results for electron impact excitation

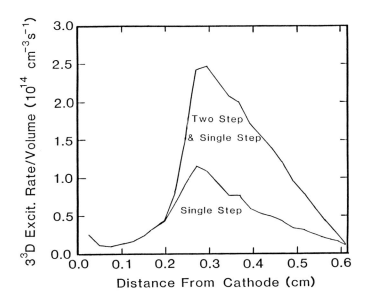

Fig. 14. Excitation rate per unit volume of He atoms to the 3^3D level as a function of distance from the cathode in the 0.846 mA/cm^2 discharge. The lower curve includes only (single step) excitation from the ground level. The upper curve includes excitation from the ground level and (two step) excitation from the 2^3S metastable level.

rates per unit volume of the singlet He levels and for ionization from the ground 1^1S level are even more symmetric about the cathode fall-negative glow boundary. Thus single-step excitation cannot explain the fact that the negative glow is much brighter than the cathode fall region.

The inclusion of two-step excitation from the 2^3S metastable level produces an excitation rate per unit volume which is very asymmetric about the cathode fall-negative glow boundary. The calculated excitation rate per unit volume in the negative glow peaks at a value 20 times greater than the minimum in the cathode fall region.

These results indicate that electron-ion recombination in the negative glow is not required to explain the greater brightness of the negative glow in comparison with the cathode fall region. We have argued in the past that if the electron temperature is significantly above gas temperature then recombination is not an important term in the ionization (particle) balance of the negative glow (Den Hartog et al. 1988). The rate per unit volume of electron collisionally stabilized electron-ion recombination scales as $(n_e^c)^3 (T_e^c)^{-4.5}$ (Deloche et al. 1976). The strong temperature dependence reduces the importance of electron stabilized recombination. Nevertheless at least some electron-ion recombination does occur in the negative glow. The observation of He_2 emission bands is convincing evidence for electron-ion recombinations (Den Hartog et al. 1988). Electron-ion recombination in He, especially dissociative recombination of He_3^+ and He_4^+, has been the subject of controversy in the past; some lingering concern remains that mechanisms may not be completely understood (Delpech et al. 1975). Our experiments and simulations indicate that electron-ion recombination is not important in the ionization balance of the negative glow, and is not needed to explain the greater brightness of the negative glow in comparison with the cathode fall region.

7. SUMMARY, CONCLUSION, AND FUTURE WORK

This investigation focused on the physics of the negative glow region in cold cathode He discharge.

The density ($n_e^c = 5 \times 10^{11}$ cm^{-3}) and temperature ($k_B T_e^c = 0.12$ eV) of the cold electrons in the negative glow of a clean He glow discharge at 0.846 mA/cm^2 and 3.50 Torr are determined using laser spectroscopy. Rates for both inelastic and superelastic collision of cold elecrons with excited atoms are measured using laser induced fluorescence and absorption spectroscopy. The density and temperature of the cold electrons are determined from these rates.

These cold electrons are trapped in a potential energy well in the negative glow. The ionization balance of the negative glow is dominated by a process analogous to ambipolar diffusion in which ions drift in a reversed field to the anode. The power balance of the cold electrons is dominated by cooling due to recoil during elastic collisions with neutral atoms, and by heating due to Coulomb collisions with hot electrons injected into the negative glow from the cathode fall region. The hot electron density ($n_e^h = 1.6 \times 10^9$ cm^{-3}) and temperature ($k_B T_e^h = 3.1$ eV) are from Monte Carlo simulations. Accurate empirical electric field maps in the cathode fall region help make the Monte Carlo simulations as realistic as possible and provide the absolute normalization through an empirical current balance at the cathode. Electric field maps are generated using optogalvanic detection of Rydberg atoms.

Two-step excitation of radiating levels (via the 2^3S metastable level) causes the greater brightness of the negative glow in comparison with the cathode fall region. Electron-ion recombination is not needed to explain the enhanced atomic He emission from the negative glow but is needed to explain the molecular He_2 emission from the negative glow. Published electron-ion recombination rate constants, when combined with the measured cold electron density and temperature, indicate that electron-ion recombination is not an important term in the ionization balance of the negative glow.

The goal of this investigation is to identify the most important physical pocesses in a typical negative glow. These processes must be included in plasma simulations such as those based on the Convective Scheme. A fully self-consistent kinetic model of the cathode fall region based on the Convective Scheme has been reported (Sommerer et al. 1989a). This cathode fall model is based on the assumption of negligible ion current at the cathode fall-negative glow boundary. It is very desirable to relax this assumption and to use the Convective Scheme through the negative glow to the anode surface. Such calculations are already showing the expected field reversal. Electrode-to-electrode models of discharge plasmas which are fully self-consistent and kinetic will provide genuine predictive capability.

REFERENCES

Alkhazov G D 1970 Zh. Tekh. Fiz. 40 97 [1970 Sov. Phys. Tech. Phys. 15 66]
Allison D C, Browne J C and Dalgarno A 1966 Proc. Phys. Soc. London 89 41
Boeuf J P and Marode E 1982 J. Phys. D: Appl. Phys 15 2169
Buckingham R A and Dalgarno A 1952 Proc. Roy. Soc. London, Ser A 213 506
Crompton R W, Elford M T and Robertson A G 1970 Aust. J. Phys. 23 667
Deloche R, Monchicourt P, Cheret M and Lambert F 1976 Phys. Rev. A 13 1140
Delpech J F, Boulmer J and Stevefelt J 1975 "Advances in Electronics and Electron Physics" ed L Marton (New York, Academic) Vol 39 p 121
Den Hartog E A 1989 Ph.D. Thesis, Univ. of Wisconsin-Madison, unpublished Ch IV
Den Hartog E A, Doughty D A and Lawler J E 1988 Phys. Rev A 38 2471
Den Hartog E A, O'Brian T R and Lawler J E 1989 Phys. Rev. Lett. 62 1500
Doughty D A, Den Hartog E A and Lawler J E 1987 Phys. Rev. Lett. 58 2668
Doughty D K and Lawler J E 1984a Appl. Phys. Lett. 45 611
Doughty D K, Salih S and Lawler J E 1984b Phys. Lett. A 103 41
Druyvesteyn M J 1937 Physica 4 669
Dubreuil B and Catherinot A 1986 Phys. Rev. A 21 188
Emeleus K G 1981 J. Phy.D: Appl. Phys. 14 2179
Fon W C, Berrington K A, Burke P G and Kingston A E 1981 J. Phys. B 14 2921

Francis G 1956 "Handbuch der Physik" ed. S Flügge (Berlin, Springer-Verlag) Vol. 22 pp53-114
Gallagher T F, Edelstein S A and Hill R M 1977 Phys. Rev. A **15** 1945
Gottscho R A, Mitchell A, Scheller G R, Chan Y-Y and Graves D B 1989 Phys. Rev. A **40** 6407
Helm H 1977 J. Phys. B: Atom. Molec. Phys. **10** 3683
Kono A and Hattori 1984 Phys. Rev. A **29** 2981
LaBahn R W and Callaway J 1969 Phys. Rev. **180** 91; 1969 Phys. Rev. **188** 520; 1970 Phys. Rev. A**2** 366
Lawler J E 1985 Phys. Rev. A **32** 2977
Long W H 1978 "Northrup Research and Technology Center Technical Report No. AFAPL-TR-79-2038" (Available as Doc. No. ADA 070 819 from the National Technical Information Service, Springfield, VA 22161)
Longmire C L 1963 "Elementary Plasma Physics" (New York, Wiley) p 204
Martin W C 1987 Phys. Rev. A **36** 3575
Moore C A, Davis G P and Gottscho R A 1984 Phys. Rev. Lett. **52** 538
Nesbet R K 1980 "Variational Methods in Electron-Atom Scattering Theory" (New York, Plenum) p 188
Phelps A V 1955 Phys. Rev. **99** 1307
Rall D L A, Sharpton F A, Schulman M B, Anderson L W, Lawler J E, Lin C C 1989 Phys. Rev. Lett. **62** 2253
Segur P, Yousfi M, Boeuf J P, Marode E, Davies A J and Evans J G 1983 "Electrical Breakdown and Discharges in Gases" ed E E Kunhardt and L H Luessen, NATO ASI Series B vol 89A (New York, Plenum) pp 331-394
Sinha S, Lin S L and Bardsley J N 1979 J. Phys. B: Atom. Molec. Phys. **12** 1613
Sommerer T J, Hitchon W N G and Lawler J E 1989a Phys. Rev. A **39** 6356
Sommerer T J, Hitchon W N G and Lawler J E 1989b Phys. Rev. Lett. **63** 2361
Vriens L and Smeets A H M 1980 Phys. Rev. A **22** 940
Warren R 1955 Phys. Rev. **98** 1650
Wellenstein H F and Robertson W W 1972 J. Chem. Phys. **56** 1072; J. Chem. Phys **56** 1077
Wiese W L, Smith M W and Glennon B M 1966 "Atomic Transition Probabilities", Natl. Bur. Stand. Ref. Data Ser., Natl. Bur. Stand. (U.S.) Circ. No 4 (Washington, D. C., U. S. GPO) Vol. 1 pp 9-15

Inst. Phys. Conf. Ser. No. 113: Section 2
Paper presented at International Meeting on Optogalvanic Spectroscopy, Glasgow, 1990

Electric field measurement in the cathode sheath of a DC glow discharge in hydrogen

C. Barbeau and J. Jolly

Laboratoire de Physique des Gaz et des Plasmas, CNRS, Université Paris-Sud, 91405 ORSAY Cedex, FRANCE

ABSTRACT: Spatially resolved electric field measurements in the cathode region of a hydrogen glow discharge are performed using polarization-dependent Stark broadening of plasma-induced emission of the Balmer lines. The large concentrations of excited atoms in the sheath provide an accurate, sensitive measure of discharge electric fields.

1. INTRODUCTION

In past years, extensive experimental investigations have been devoted to the measurement of electric field in the cathode sheath region of glow discharges. The interest in cathode sheath mecanisms, mostly governed by the rapidly changing E/N quantity (ratio of electic field to molecular concentration), comes from the fact that the electron and ion kinetics and the electrical power dissipation in glow discharges are primarily determined by the sheath properties. The understanding of glow discharges is of interest because of their practical applications in plasma processing. The previously reported experimental techniques rely on the electric-field-induced breakdown of selection rules of laser-induced optical transitions (Drullinger et al 1975, Moore et al 1984, Derouard and Sadeghi 1986), the Stark splitting and/or line broadening of laser-induced high Rydberg states (Doughty and Lawler 1984, Doughty et al 1984, Ganguly and Garscadden 1985, Shoemaker et al 1988) or electron beam probe techniques (Warren 1955). The first technique has been demonstrated for plasmas containing either BCl radicals or NaK molecules and the second one has been used in rare gas discharges.

In this paper we report a method for in-situ, non intrusive measurement of the axial electric field intensity in the cathode fall region of hydrogen glow discharges. The technique, based on polarization-dependent Stark broadening of plasma-induced emission of the Balmer lines, has a dynamic range from a several tens of volts/cm to a few kilovolts/cm.

2. EXPERIMENTAL SETUP

The plasma is created in a conventional parallel-plate discharge structure with iron electrodes of diameter 30 mm. The electrodes separated by 30 mm are enclosed in a much larger stainless steal vacuum system. The discharge operates in a hydrogen flow (40 to 150 sccm) for pressures ranging from 0.1 to 1.2 Torr (~ 13 to 160 Pa). The discharge voltage is supplied by a stabilized DC power generator through a 76 kΩ load resistor.

© 1991 IOP Publishing Ltd

The spatially resolved Balmer-line emission profile is observed, through a polarizer, perpendicularly to the discharge axis. The light emitted by the plasma is first collimated and then focused onto the entrance slit of a 1.5 m focal length scanning monochromator, operating in the 2nd or 3rd order, providing a spectral resolution better than 0.02 nm. The optical system gives a spatial resolution, in a plane perpendicular to the optical axis, which depends on monochromator slit and optical aperture settings. In the present experiment a resolution better than 0.5 mm was achieved. The axis of the optical system is precisely movable so that the inter-electrode space can be investigated as a function of distance from the cathode.

3. METHOD

It has been reported previously (Benesch and Li 1984, Cappelli et al 1985, Li Ayers and Benesch 1988), that hydrogen excited atoms may have, in the cathode fall region, high kinetic energies (up to 1 keV in our experimental conditions). In a recent publication (Barbeau and Jolly 1990), we interpreted quantitatively this translational energy as resulting from gas-phase charge-exchange collisions of hydrogen ions (H^+, H_2^+, H_3^+) with H and H_2, and backscattering of hydrogen ions on the cathode surface. When observing the plasma perpendicularly to the discharge axis, it is important to note that the excited atoms which are observed as high velocity atoms (the atoms having an important velocity component in the direction of observation) are created, in major part, by the latter mechanism. In the present experimental conditions, about 60% of the hydrogen ions reaching the cathode are backscattered, H_2^+ and H_3^+ ions are dissociated and the reflected particles, mostly neutral atoms, carry an average of ~ 40% of the total energy of the incident ion beam. Moreover the maximum emission of the Balmer lines observed close to the cathode could not be explained by electronic excitation of the fast atoms in the fundamental state. Thus, according to McCracken and Freeman (1969) and McCracken and Erents (1970), we had to consider that most of the Balmer line emission results from energetic atoms which escape the electrode in an excited state, and then radiate when travelling along the discharge. The emission profiles of the atomic lines in the cathode fall region thus need to be analysed taking into account both the Doppler broadening and shifting resulting from these energetic atoms and the Stark effect induced by the local electric field.

We made the assumption that the Doppler components of the different Balmer lines are identical, relatively, in terms of $\Delta\lambda/\lambda$ and depend only on the distance from the cathode. This essentially implies that the production and relaxation processes of the excited atoms $H^*(n)$ lead to an energy distribution that is independent of the principal quantum mumber n. This approximation could be discussed in a more detailed analysis of the differents components of the Doppler profiles of the Balmer lines as it was previously done for $H\alpha$ (Barbeau and Jolly 1990). Nevertheless the good quality of the fits presented further in this paper confirms experimentally the validity of this assumption in the present experimental conditions.

Since the Stark effect on the $H\alpha$ line may be neglected, with a good approximation, for electric field values lower than 2 kV/cm, the knowledge of the local Doppler parameters may be directly obtained from the observation of the $H\alpha$ (λ_0 = 656.3 nm) emission lineshape. It is therefore possible to calculate the synthetic emission profiles of the other Balmer lines if the intensity distribution of the Stark manifold corresponding to the local electric field value is known.

The theoretical Stark splitting of the hydrogen atom levels n = 2 to 6 was computed using the method described by Lüders (1951). For the electric field intensities considered in this study, the splittings of the fine structure and Stark effect are in the same order of magnitude for the n = 2 level. This also applies to the n = 6 level if we consider the possibility of measuring low electric field values (E < 100 V/cm). Therefore, the linear Stark theory is no longer valid. The hamiltonians of the fine structure and Stark effect are treated with a first order perturbation procedure which has as its starting point the unperturbed eigenfunctions.

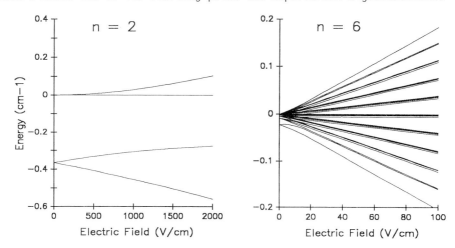

Fig. 1. Calculated Stark splitting on the fine structure of the n=2 and n=6 levels of atomic hydrogen vs electric field.

The splitting of the n = 2 and n = 6 levels are shown in figure 1. The Stark manifold of the n = 6 level is presented at low electric field values for which the non-linear Stark effect is effective. The Lamb shift as well as the hyperfine structure are neglected. The relative intensities of the π-components ($\Delta m_j = 0$) of the Balmer lines $H\alpha$, $H\beta$, $H\gamma$ and $H\delta$ were calculated using the dipolar electric approximation for different field intensities (Bethe and Salpeter 1977).

Whereas the Stark effect increases with the principal quantum number n, the Balmer line emission decreases rapidly with increasing n. The $H\delta$ transition (n=6 $\rightarrow$ n=2, λ_0 = 410.2 nm) was chosen as a good compromise between the optical transition intensity and electric field sensitivity. The synthetic profiles of the $H\delta$ line, at different positions along the discharge axis, may be calculated from the convolution of the Stark pattern of the $H\delta$ transition with the local $H\alpha$ emission lineshape. An example is presented in figure 2, for an electric field value of 1200 V/cm.

4. EXPERIMENTAL RESULTS

The $H\alpha$ and $H\delta$ profiles are recorded using different values of the monochromator spectral resolution so as to keep constant the convolution product of the Doppler function and apparatus function (i.e. the convolution product remains the same function of the parameter $\Delta\lambda/\lambda$ for both $H\alpha$ and $H\delta$ recordings). Other causes of line broadening such as the Stark effect due to space charge field are negligible in the present

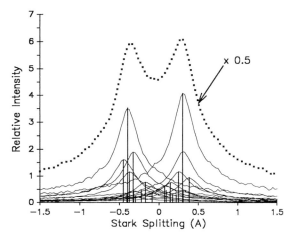

Fig. 2. Synthetic profile of the Hδ ($\Delta m_j = 0$) transition (dotted line) calculated for an electric field value of 1200 V/cm.

conditions. The local electric field intensity in the cathode sheath can therefore be deduced by choosing the best fit of the experimental Hδ lineshape with synthetic profiles calculated for different values of the electric field. The fit allows us to determine the field value with a precision of $\sim$ 50 V/cm. Typical results, for p = 0.6 Torr, V = 900 V and i = 6 mA, obtained at locations 3 mm and 10 mm from the cathode, are shown in figure 3. The agreement between experimental and calculated profiles is very good.

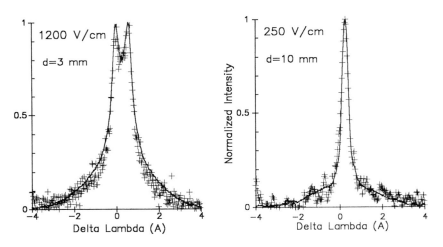

Fig. 3. Measured (+) and calculated (full curve) normalized intensity profiles of the Hδ line. Electric field values of 1200 and 250 V/cm gave the best fits respectively at positions 3 and 10 mm from the cathode surface. The discharge current was 6 mA, the pressure 0.6 Torr and the voltage drop 900 V.

The entire cathode fall region was investigated. Figure 4 shows the spatial variations of the electric field as a function of distance from the cathode for different gas pressures. For the higher pressure values (0.6 to 1 Torr), the electric field profile is linear over most of the cathode fall region, except at both ends. But when the pressure decreases, a non-linearity appears near the negative glow boundary. This effect was previously observed, at least qualitatively, by Warren (1955). The integral of the experimental electric field profile corresponds to the applied voltage, within a few percent, for all the discharge conditions used in this study. The precision on the results is limited by both the uncertainties on the axial position, mostly induced by a possible parallelism defect between the cathode surface and the optical axis, and the limited spatial resolution (0.5 mm) resulting in an average electric field value measured at the location of the optical beam in the sheath. The dynamic range of the method is, at present, limited both for the high field values by the approximation of a negligible Stark effect on the H_α line and for the low fields by the difference of the fine structure values of the H_α and H_δ line. Improvements based on a more sophisticated analysis of the profiles are under progress.

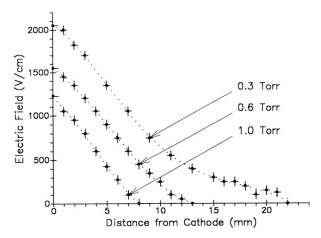

Fig. 4. Sheath electric field as a function of distance from the cathode for a discharge current of 6 mA and different values of the gas pressure. The size of the crosses represents the uncertainties on both the electric field measurement and axial position.

In summary, we present for the first time, an experimental determination of the cathode fall electric field value using plasma-induced emission in a hydrogen glow discharge. This non-intrusive method has the advantage, among others, to be performed without the use of sophisticated and expensive apparatus (laser, electron beam ...) and provides, nevertheless, an accurate measure of the absolute electric field intensity in the cathode fall region. This method could be extended to RF discharges by using a time-resolved spectroscopic detection of the atomic lines, as well as in gas mixtures containing hydrogen provided that the Balmer line emission is sufficiently important in the cathode sheath. The transition used to evaluate the electric field has to be chosen, in each case, according to the field strengh and light intensity.

References.

Barbeau C and Jolly J 1990 to be published in *J. Phys.* D **23**
Benesch W and Li E 1984 *Opt. Lett.* **9** 338
Bethe H A and Salpeter E E 1977 *Quantum Mechanics of One- and Two-Electron Atoms*, (New York: Plenum/Rosetta)
Cappelli A, Gottscho R A and Miller T A 1985 *Plasma Chem. Plasma Proc.* **5** 317
Derouard J and Sadeghi N 1986 *Opt. Commun.* **57** 239
Doughty D K and Lawler J E 1984 *Appl. Phys. Lett.* **45** 611
Doughty D K, Salih S and Lawler J E 1984 *Phys. Lett.* **A 103** 41
Drullinger R E, Hessel M M and Smith E W 1975 *Laser Spectroscopy*, ed S Haroche, J C Pebay-Peyroula, T W Hänsch and S E Harris, (Berlin: Springer Verlag) p.91
Ganguly B N and Garscadden A 1985 *Phys. Rev.* A **32** 2544
Li Ayers E and Benesch W 1988 *Phys. Rev.* A **37** 194
Lüders G 1951 *Ann. Phys.* **8** 301
McCracken G M and Freeman N J 1969 *J. Phys.* B **2** 661
McCracken G M and Erents S K 1970 *Phys. Lett.* **31** A 429
Moore C A, Davis G P and Gottscho R A 1984 *Phys. Rev. Lett.* **52** 538
Shoemaker J R, Ganguly B N and Garscadden A 1988 *Appl. Phys. Lett.* **52** 2019
Warren R 1955 *Phys. Rev.* **98** 1650

Optogalvanic effects in very low current discharges

R.J.M.M.Snijkers, G.M.W Kroesen, J.M.Freriks, F.J. de Hoog.

Faculty of Technical Physics, Eindhoven University of Technology,
POB 513, 5600 MB Eindhoven, the Netherlands

ABSTRACT: Usually the optogalvanic effect is ascribed to a disturbance in the ionisation–excitation equilibrium in a discharge plasma through the resonant absorption of radiation. Modelling of this phenomenon is a complex task: the absorption of the radiation, the change in population of excited and ionised states and the translation to external discharge parameters have to be calculated. In a discharge without space charge modelling becomes less complicated. We carried out experiments in a Townsend discharge at current densities of about 10^{-8} A/cm^2 and found optogalvanic effects which could only be interpreted in terms of secondary ionisation. Observed zero–crossings when varying the position of the irradiated volume are discussed. The results of the experiments illustrate the non–diffusive character of the transport of resonance radiation.

1. INTRODUCTION

Generally speaking any effect on current and/or voltage in an electronic device by the resonant or non–resonant absorption of radiation from the optical spectrum may be characterized as an optogalvanic effect. But in practice this name has been used only for the change in voltage and/or current when radiation is absorbed in active discharge devices. Among the first to describe this effect was Penning (1928). He measured breakdown potentials in discharges between parallel electrodes of neon with 0.001 % of argon. The discharge was irradiated by light from a positive column discharge in pure neon. When irradiated a considerable increase of the breakdown potential occurred. His explanation for the effect, that he called a new photoelectric effect, was the decrease of the ionisation of argon by neon metastables through the destruction of the latter by what we call now optical pumping. It may be clear that this experiment was part of the evidence for the effect that now bears his name.
The change in the ionisation–excitation equilibrium brought about by the absorption of radiation causes measurable changes in the external parameters like voltage and current. The development of tuneable dye lasers has increased the interest in and possibilities of the optogalvanic effect (OGE) considerably. The effect is easy to measure compared with other diagnostic methods and usually the direct electrical detection yields good signal to noise ratios. Therefore the OGE has become a valuable tool for spectroscopy of discharges, but quantitative modelling i.e. relating measured current and voltage changes with the characteristics of the incoming radiation can still meet considerable problems.
In fact a system of coupled rate equations for all pertinent electronic states of the parent gas molecules should be solved. To the regular radiative–collisional coupling, the coupling by the absorption of the incoming radiation should be added. Next to that the Boltzmann equation for electrons and ions including transport should be

© 1991 IOP Publishing Ltd

solved. This also necessitates a selfconsistent solution of Poisson's equation for the electric fields. Finally from all these data and from the external circuit parameters the values for current and/or voltage changes should be calculated.
Because the OGE usually is small compared with the discharge current and/or voltage, a number of investigators (Pepper 1978, Lawler 1980) have succeeded to simplify the above program and still find good quantitative values for the OGE. A recent example of this kind of modelling is by Stewart et al. (1990), who studied effects in the neon positive column.
When irradiating strong inhomogeneous parts of a discharge e.g. the transition from cathode dark space to glow, interesting spectroscopic effects with strong impact on discharge diagnostics can be observed (den Hartog et al 1988), but quantitative discharge modelling becomes more complicated. It is known that in hollow cathode discharges in neon the effects with can be observed show e.g. change of sign when scanning the cathode fall (van Veldhuizen et al. 1984). In the same work also a dependency of the OGE from the radiation frequency within the absorption line has been reported.
In order to study these effects and to circumvent the complicated interpretation necessary, it looks worthwhile to observe these effects in very low current selfsustaining discharges between parallel plates. In such discharges the collisional radiative processes are well defined and the electrode structure leads to a one–dimensional approach in the modelling. Excited state densities as well as electron densities are so small that interaction of excited level populations through electrons is absent. Also heating of the electron gas by superelastic collisions or ionisation through collisions of two metastables may be ruled out.
These suggestions have been reported by Kravis and Haydon (1981). They reported on observation of OGE signals in a prebreakdown non–selfsustaining Townsend discharge in neon brought about by the radiation from a pulsed dye laser. One of their conclusions is that the incidence of high energy photons on the cathode arising from the depletion of the metastable levels by optical pumping should be considered in the modelling of any optogalvanic effect.
That the effect in a Townsend discharge may be considerable has already been shown by Penning (1928), but can also be seen from the following reasoning.
If one assumes that the absorbed power P will lead to a decrease of the electrical power fed into the discharge, from

$$I_0 V_0 = (I_0 + \delta I_0)(V_0 + \delta V_0) + P,$$

where I_0 and V_0 are the current and voltage of the discharge without irradiation, it can be easily seen that in a constant current situation

$$\delta V = -\frac{P}{I_0}.$$

So the signal is large for smaller discharge currents.

In this paper we describe an experiment where the optogalvanic response of a Townsend discharge in neon is measured as a function of the position of irradiation. In 2 the pertinent parameters of a Townsend discharge are reviewed including a derivation of the effect that can be expected. In 3 the experiment is described and the experimental results are given. In 4 the results are given of a study of the balance of arrival at the cathode of metastables and high energy photons which will prove to be essential for the interpretation of our results. Here we will calculate the optogalvanic effect and compare the results with the experiment.

2 THE OPTOGALVANIC RESPONSE OF A TOWNSEND DISCHARGE

Since in Townsend discharges between flat parallel circular electrodes the current densities are of the order of 10^{-8} A/cm^2, space charge can be neglected and the

electric field within the discharge volume is homogeneous. This enables one to assume that the electron energy distribution is in equilibrium with the field and to describe the discharge according to the well known Townsend model.
The electron density $n_e(x)$ grows starting from the cathode value $n_e(0)$ according to

$$n_e(x) = n_e(0) \exp(\alpha x),$$

with α, the primary ionisation coefficient fulfilling

$$\alpha = \frac{k^{ion} N_0}{v_e}.$$

Here k^{ion} is the electron ionisation rate, N_0 the gas density and v_e the electron drift velocity. For the ion density $n_i(x)$ follows

$$n_i(x) = \frac{v_e}{v_i} n_e(0) \{\exp(\alpha d) - \exp(\alpha x)\},$$

where v_i is the ion drift velocity and d is the electrode distance.
Under the experimental conditions of our experiment (a reduced pressure of 3.7 mmHg, a discharge voltage V of 237 volts and a current density of $4.0 \; 10^{-8}$ A/cm^2) the ion density at the cathode amounts to approximately $2 \; 10^6$ cm^{-3} and the electron density at the anode to appr. $2 \; 10^4$ cm^{-3}. In this case all the remarks made in the introduction about the conditions for simple discharge modelling are fulfilled. This means that ionisation takes place only through direct impact of electrons with ground state atoms and that optogalvanic effects can only work through secondary ionisation, which takes place at the cathode. To implement secondary ionisation in a self–sustaining discharge it is necessary to discuss the expression for the current amplification in a non–selfsustaining discharge of the current density $j(d)$ with distance assuming a constant current source at the cathode j_0 which reads

$$\frac{j(d)}{j_0} = \frac{\exp(\alpha d)}{1 - \Gamma\{\exp(\alpha d) - 1\}}.$$

From the denominator the condition for selfsustainment of the discharge is

$$1 - \Gamma\{\exp(\alpha d) - 1\} = 0, \tag{2.1}$$

with Γ to be written as

$$\gamma_i + \gamma_m f_m \frac{\alpha_m}{\alpha} + \gamma_p f_p \frac{\alpha_p}{\alpha}. \tag{2.2}$$

Here γ_i, γ_m and γ_p are the secondary electron emission coefficients of ions, metastable and resonant photons upon arrival at the cathode resp.; α_m and α_p are the number of metastable and resonant states respectively formed by an electron along a unit length in the direction of the field. The fraction of the total amount of metastable states formed in the discharge arriving at the cathode is represented by f_m, whereas the same fraction of resonant photons arriving at the cathode is f_p. In fact working in neon one has to realise that there will be four s–levels active in secondary ionisation and that expression (2.2) should contain 5 terms. For the sake of our reasoning we will confine ourselves here to a model with only one metastable state and one resonant state. In neon these are the $^3P_0$–state, which is metastable and the $^3P_1$–state which is resonant. It will be shown at the end of this section that the model can be easily expanded to a second metastable state and a second resonant state.
Irradiating part of the discharge and pumping the $^3P_0$–state to the $^3P_1$–state through 588 nm radiation from a tuneable dye laser, all constants in expression

(2.1) and (2.2) remain the same except for the fractions f_m and f_p. To find the relation between the change in discharge voltage δV and the changes δf_m and δf_p we write expression (2.1) as

$$V = \frac{1}{\eta} \ln(1 + \frac{1}{\Gamma}), \qquad (2.3)$$

with η the number of ionisations by an electron along the traverse of a unit of potential difference. From this it follows that

$$\delta V = -\frac{1}{\eta} \frac{1}{(\Gamma^2 + \Gamma)} (\gamma_m \, \delta f_m \frac{\alpha_m}{\alpha} + \gamma_p \, \delta f_p \frac{\alpha_p}{\alpha}). \qquad (2.4)$$

It appears that the expression for f_m can be found from the solution of the diffusion equation for the metastable state density

$$\frac{d^2 n_m}{dx^2} = -\frac{n_m}{\tau} + \alpha_m \, v_e \, n_e(0) \exp(\alpha x) \qquad (2.5)$$

with the boundary conditions

$$n_m(0) = 0 \qquad n_m(d) = 0. \qquad (2.6)$$

The factor τ represents the lifetime of a diffusing metastable before undergoing a collisional transition to another state. Under our experimental conditions for a pressure of 3.7 mmHg from Phelps' data (Phelps 1959) it can be shown that the lifetime of a metastable state is determined by the diffusion process and that collisional transitions may be neglected. In that case the calculation of f_m becomes simple. From the Green function solution of this simple diffusion problem one can show that the probability for a metastable created at position x to arrive at the cathode is

$$p_c(x; d) = -\frac{x}{d} + 1.$$

Then by integration of $\alpha_m \, v_e \, n_e(0) \exp(\alpha x)$ weighted by p_c over the discharge volume it follows that

$$f_m = \frac{1}{\alpha d} - \frac{1}{\exp(\alpha d) - 1}.$$

If one assumes that a plane at the position $x = \xi$, with $0 < \xi < d$, also absorbs the metastables, boundary conditions (2.6) should be extended with

$$n_m(\xi) = 0.$$

In that case

$$f_m(\xi) = \frac{1}{\alpha \xi} \frac{\exp(\alpha \xi) - 1}{\exp(\alpha d) - 1} - \frac{1}{\exp(\alpha d) - 1}, \qquad (2.7)$$

so

$$\delta f_m(\xi) = -\frac{1}{(\alpha d)} + \frac{1}{\alpha \xi} \frac{\exp(\alpha \xi) - 1}{\exp(\alpha d) - 1}. \qquad (2.8)$$

Using the expression for p_c and assuming that all the metastables pumped away at the plane $x = \xi$ will emerge as resonant states it is possible to calculate $\delta f_p(\xi)$ if one has a proper model of the transport of the resonance radiation emitted from the $^3P_1$–state. Assuming pure diffusion of the resonant photons one can show that

$$\delta f_p(\xi) = -\delta f_m(\xi) \frac{\alpha_m}{\alpha_p}. \qquad (2.9)$$

The proof can be found by using the fact that the metastable current lost to $x = \xi$ from parts of the discharge with $0 < x < \xi$ is again the integration of $\alpha_m v_e n_e(0) \exp(\alpha x)$ between $x = 0$ and $x = \xi$, but now weighted by

$$p_a(x;\xi) = 1 - p_c(x;\xi) = \frac{x}{\xi},$$

and that the loss of metastables at the plane ξ from parts $\xi < x < d$ follows from the integration of $\alpha_m v_e n_e(0) \exp(\alpha x)$ between $x = \xi$ and $x = d$ weighted with

$$p_c(x-\xi\,;d-\xi) = \frac{d-x}{d-\xi}.$$

The expression (2.8) becomes simple because the f's are only dependent on α, d and ξ, and not on the diffusion coefficients. In the case of non–diffusive transport of the resonance radiation expression (2.9) can be written:

$$\delta f_p(\xi) = -\delta f_m(\xi) \frac{\alpha_m}{\alpha_p} \epsilon(\xi), \qquad (2.10)$$

which means that a photon emerging from the plane ξ has a probability to be lost at the cathode of

$$\epsilon(\xi) \left(-\frac{\xi}{d} + 1\right).$$

The change in the discharge voltage δV can now be written

$$\delta V = -\frac{1}{\eta} \frac{1}{(\Gamma^2 + \Gamma)} \delta f_m(\xi) \frac{\alpha_m}{\alpha} \{\gamma_m - \gamma_p \epsilon(\xi)\}. \qquad (2.11)$$

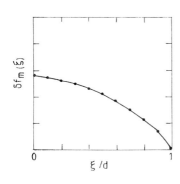

Fig.1 Change in fraction of metastables arriving at cathode $\delta f_m(\xi)$ against position of irradiated volume.

If the discharge characteristics are known, δV can be calculated. It is clear that $\delta f_m(\xi)$ is a positive function of ξ. In figure 1 $\delta f_m(\xi)$ is shown for a discharge under the same experimental conditions as our experiment. For pure diffusive transport of resonance radiation $\epsilon(\xi) = 1$ for all ξ. So in that case no change of sign of the OGE can occur. Since in our experiment such a change occurs this is proof of the non–diffusive character of the high energy photon transport.

In the excitation scheme of neon four levels play a role. If the probability to pump the $^3P_2$–state to the $^3P_1$–state, the $^3P_0$–state and the $^1P_1$–state are F_{p1}, F_{m2} and F_{p2} respectively and $F_{p1} + F_{m2} + F_{p2} = 1$, it follows that

$$\delta V = -\frac{1}{\eta} \frac{1}{(\Gamma^2 + \Gamma)} \delta f_m(\xi) \frac{\alpha_{m1}}{\alpha} \{\gamma_m (1 - F_{m2}) - \gamma_p (F_{p1} + F_{p2}) \epsilon(\xi)\}.$$

Here it was assumed that both metastable states have the same γ_m and both resonant photons have the same γ_p. Also it was assumed that $\epsilon(\xi)$ is the same for both kinds of photons. If these assumptions are not valid the expression for δV becomes more complicated but essential features concerning the possibility of sign change are not lost.

3 EXPERIMENT

As already mentioned in section 2 the experiment was carried out in a Townsend discharge in neon carrying a current of $0.5 \cdot 10^{-6}$ A and having a burning potential of 237 volt at a reduced pressure of 3.7 mmHg. The electrode distance was 11.0 mm and the electrode radii were 20 mm. The gas was spectroscopically pure neon.

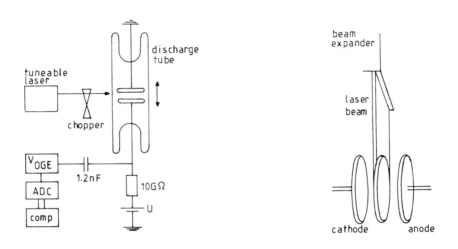

Fig.2 Schematics of OGE experiment.

Fig.3 Geometrical set up of irradiated volume in discharge

Comparison with data of breakdown potentials by Frouws (1957) of pure neon led us to the conclusion that impurities did not take part in the ionisation process. The laser radiation was generated by a cw ring dye laser pumped by an argon ion laser (Spectra Physics 380D,171). The linewidth and thus the wavelength stability amounted to 10 Mhz. The laser beam was expanded in such a way that within the discharge a slablike volume with a width of 1 mm parallel to the electrodes was irradiated. The optogalvanic signals were measured at the anode of the tube. As a buffer use was made of a high voltage mosfet used as a source follower. The gate current was about 1 nA. The experiment is schematically drawn in figure 2. In figure 3 the geometrical set up of the irradiation is sketched.

When measuring the effect it is of importance to estimate the probability for metastable $^3P_2$ states present within the irradiated volume by excitation or diffusion to be pumped to the $2p_2$-state. The laser line usually is positioned in the centre of the absorption line by maximizing the response. The absorption width of the line is dominated by Doppler broadening amounting to about 1.4 Ghz. So not every $^3P_2$-state will interact with the laser radiation. Even if we take into account the power broadening of the absorption at laser intensities of 25 mW/cm^2 which is about 50 Mhz, by far not all metastable states within the irradiated volume interact with the radiation field. Experiments show however a saturation effect indicating that optical pumping in our case is effective. In figure 4 the OGE measured when the centre of the irradiated volume is 1 mm from the cathode is plotted against laser beam power.

When one realizes that the lifetime of a metastable within the irradiated volume is determined by diffusion and is approximately 10^{-4} s, with an elastic scattering frequency of about 14 Mhz the $^3P_2$–state diffuses through velocity space and will during its stay in the irradiated volume of real space pass through the irradiated volume of the velocity space after an average of about 36 elastic collisions. It will then get in interaction with the radiation field and within a time of about 100 ns get into the upper level. This time is comparable with the elastic scattering time. As substantiated by the results of figure 4 we may conclude from the above consideration that indeed all metastables ever to be present in the irradiated volume are optically pumped to the $2p_2$–state from which they will decay to the other s–states.

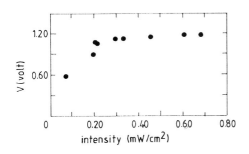

Fig.4 *Optogalvanic effect when irradiating at 1 mm from cathode as a function of laser intensity of the 588.2 nm line.*

So the assumptions made in section 2 on the disappearance of metastables at the position $x = \xi$ leading to the expression (2.10) are valid in the case of the experiments.

The experiments have been carried out with a laser intensity of 40 mW/cm² at a wavelength of 588.2 nm. The laser beam was chopped with a frequency of 30 Hz

In figure 5 the OGE as a function of time has been plotted at various values for ξ.

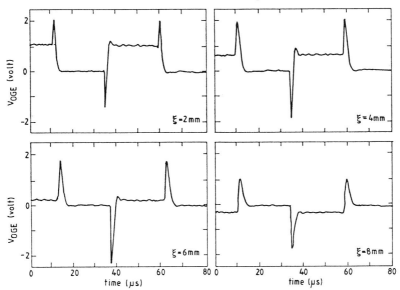

Fig.5 *Optogalvanic effect as a function of time at various positions of the irradiated volume for a wavelength of 588.2 nm.*

As indicative for the value of δV the plateau was taken. Up to now no analysis of the transients at the chopping times was undertaken, but from work with pulsed dye lasers (Erez et al. 1979) it appears that these transients can be interpreted as containing the time constants for the relaxation of both resonant states and metastable states. In figure 6 the OGE has been plotted as a function of the position of the irradiated volume within the discharge at the experimental conditions discussed above. Not only the response to 588.2 nm radiation is given but also the response to radiation of 603.0 nm. This radiation is able to optically

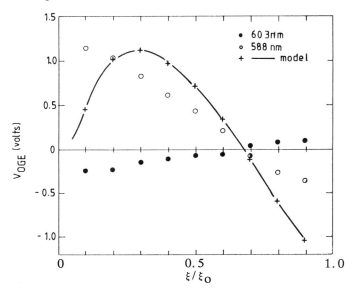

Fig.6 Static optogalvanic effect in neon as a function of the irradiated volume.

pump the $^3P_1$–state through the $2p_2$–state to the $^3P_2$–state. The response to the 588.2 nm light shows the predicted change of sign indicating the non–diffusive character of the transport of the 74.3 nm photons. The response to the 603.0 nm irradiation shows the inverted effect which is qualitatively easy to explain since because of the non–diffuse character of the transport of the 74.3 nm photons a photon state converted to a metastable has a larger probability to arrive at the cathode.

4 DISCUSSION OF RESULTS

In order to explain the change of sign as well as the absolute value of the experimental data, it is necessary to get an idea about the values of $\epsilon(x)$. To this end a Monte Carlo calculation was carried out in which 1000 photons with a wavelength of 74.3 nm were emitted at the position x between the electrodes. This was repeated for various other values of x. It was asssumed that the absorption of these photons was governed by a Doppler broadened absorption line at room temperature. Both the effects of the ^{20}Ne and the ^{22}Ne isotopes were taken into account. Also the reflection coefficient at the electrodes was taken to be 0.1. At a reduced pressure of 3.7 mmHg, to find the balance between reasonable statistics and computer time, a maximum amount of scattering events per photon of 20000 was taken. In that case only 6 % of the photons had to be aborted because they did not reach the electrodes. From the number to arrive at either the cathode or the anode the probability $p_c(x;d)$ for photons to arrive at the cathode could be derived. In figure 7 the result of the calculation is shown. The probability for a photon to

arrive at the cathode under the conditions of the experiment only deviates a few percent from $p_c(x;d)$ for metastables. With the values from this calculation it is possible to calculate the value of δV.
To this end values of η, γ_i, γ_m, γ_p, α_m/α, α_p/α, f_m and f_p are necessary. Since the reduced field strength is known to be 58.2 V/cm mmHg, the value for the primary ionisation coefficient η is 0.12 V^{-1}. The overall second ionisation coefficient Γ is then from (2.1) 0.55. The maximum value of $\delta f_m(\xi)$ for $\xi = 0$ amounts to 0.28, which is also the value for f_m and is supposed to be the value for f_p. From the experiments and the calculations of $\epsilon(x)$, considering the zero crossing point ξ_0 of the experiment it follows

$$\epsilon(\xi_0) = \frac{\gamma_m}{\gamma_p} = 1.06.$$

Also from data by Druijvesteijn and Penning (1940) at an E/p_0 value of 58.2 V/cm mmHg it seems reasonable to assume $\alpha_m/\alpha = \alpha_p/\alpha = 1$. Finally assuming $\gamma_i = \gamma_m$, one finds for these coefficients a value of 0.035.

Now it is possible to calculate δV from (2.11). In figure 6 the results of this calculation are shown. It is clear that, in spite of a number of uncertainties concerning the above assumptions, the model explains the observation quite well.

As long as the various values for the discharge coefficients are not known with greater accuracy, it is not very sensible to refine the Monte Carlo calculations e.g. by taking into account pressure broadening through a Voigt profile. One should keep in mind that the deviations at larger values of ξ are related to the assumption in the model that the electrodes are infinitely large.

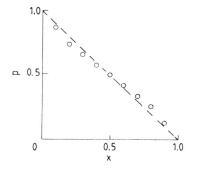

Fig. 7 The probability of arrival at the cathode of a 74.3 nm photon emitted at position x at a pressure of 3.7 mmHg.

An interesting point still to be discussed is the response to the 603.0 nm line. It is now not possible to find a simple expression like (2.10) to relate δf_p and δf_m. It may be clear however that the expression for δV will now not scale with α_m, but with α_p, whereas the factor within parenthesis, indicating the unbalance, will be of the same order of magnitude but of the opposite sign. These results fit in this respect well within the framework of the model.

CONCLUSION

It was shown that optogalvanic effects which occur in Townsend discharges are due to secondary ionisation, i.e. through the production of electrons at the cathode of the discharge. The simplicity of the discharge model makes it possible to predict quantitative values of voltage variations from the selfsustainment condition.
A change of sign of the optogalvanic effect, observed in these discharges when varying the irradiated volume, was ascribed to the difference in transport characteristics of the various types of excitation energy. The metastables should show purely diffusive behaviour, whereas the transport of resonance radiation shows deviations from diffusive behaviour. Monte Carlo calculations are able to give an indication of the difference, which was confirmed in the experimental results.
These experiments have shown that in modelling the OGE in cathode falls and transitions between cathode fall and negative glow the balance between arrival of metastables and resonant photons at the cathode should be taken into account.
Thus the study of the optogalvanic effect in these simple discharges can contribute to the experimental study of the transport of resonance radiation.

REFERENCES

Druijvesteijn M J and Penning F M 1940 *Rev. Mod. Phys.* **12** 87
den Hartog E A, Doughty D A and Lawler J E 1988 *Phys. Rev.* **A38** 2471
Erez G, Lavi S and Miron E 1979 *IEEE J Quantum Electron.* **15** 1328
Frouws S M 1957 *Proc. 3rd Int. Conf. on Phenomena in Ionised Gases, Venice* (Amsterdam: North Holland) p 341
Lawler J E 1980 *Phys. Rev.* **A22** 1025
Kravis S P and Haydon S C 1981 *J. Phys D: Appl. Phys.* **14** 151
Penning F M 1928 *Ned. Tijdschr. voor Natuurkunde* 137
Pepper D M 1978 *IEEE J Quantum Electron.* **14** 971
Phelps A V 1959 *Phys. Rev.* **114** 1011
Stewart R S, McKnight K W and Hamad K I 1990 *J. Phys. D: Appl. Phys.* **23** 832
van Veldhuizen E M, de Hoog F J and Schram D C 1984 *J. Appl. Phys* **56** 2047

Optogalvanic detection of excited-state photoionization in the neon positive column

J. Halewood, R.C. Greenhow
Department of Physics, University of York, Heslington, York Y01 5DD

Abstract: Non-resonant optogalvanic signals have been detected in neon positive column discharges. These signals are attributed to photoionization of excited states of atomic neon. A theory is developed to quantify these signals and the possibility of obtaining state-selective photoionization cross-sections is discussed.

In addition to the usual optogalvanic resonance signals we have observed continuum signals in neon positive column discharges at wavelengths between 300 and 610 nm. These signals are several orders of magnitude smaller than typical OG resonance signals but are still detectable in suitably quiet discharges. The signals have a negative characteristic when detected at the cathode, i.e. they show an increase in current. They are also found to be linear with intensity (eventually saturating at high intensities), indicating a single rather than multiphoton process. These signals can be attributed to continuum absorption mechanisms within the discharge plasma. The possible mechanisms considered are those of inverse bremsstrahlung and photoionization; scattering processes are not considered as they are too weak to account for the size of the observed signals. Autoionization, not strictly a continuum process, is also implicated at some of the wavelengths involved.

Figure 1 shows the continuum signals observed in a 0.5 torr neon discharge using Ar^+-ion, R6G-dye laser and Hg discharge lamp sources. The signals were detected with a Brookdeal 9501E lock-in amplifier with a time constant of 30 sec. The background noise of the discharge was approximately 5 pA compared to the theoretical shot noise limit of 2.3 pA, which is taken to indicate discharge purity. The signals increase with wavelength up to 475 nm., which approximately corresponds to the threshold for photoionization of the $2p^53p$ manifold, becoming steadily smaller at longer wavelengths. Continuum signals have also been observed in a He-Ne discharge [Stark 86] and have a very different wavelength dependence, showing a roughly linear increase with increasing wavelength. The difficulties of modelling a He-Ne gas mixture prevents a theory being proposed for that system.

© 1991 IOP Publishing Ltd

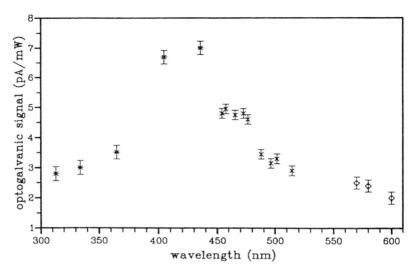

Figure 1: continuum signals in 1/2 torr discharge: ∗:Hg lamp,×:Ar$^+$-ion,◊:dye laser data

Models of Continuum Absorption: Inverse bremsstrahlung
There are two forms of inverse bremsstrahlung taking place within a plasma, arising from electron-ion and electron-atom interactions. The ratio of the absorption rates for the two two processes is given by [Bekefi 66]

$$\frac{\eta_{ei}}{\eta_{ea}} \simeq 6.5 \times 10^{-6} \frac{\overline{G}(T,\omega)Z^2}{bT^{1.5}} \frac{N_{ion}}{N_{atom}} \qquad (1)$$

where $\overline{G}(T,\omega)$ is the Gaunt factor and b is derived from the electron-atom collisional frequency $\nu = bN_{atom}$. In the neon positive column studied here $N_{ion}/N_{atom} \simeq 1 \times 10^{-7}$ and the electron-atom bremsstrahlung is expected to dominate by over five orders of magnitude at optical frequencies.

For electron-atom bremsstrahlung the differential absorption rate at high frequencies is given by Holstein [Holstein 65] as

$$\eta_\omega(\nu) \simeq \frac{N_{atom}e^2}{3\epsilon_o c\hbar\omega^3}(v')^3(1 - \frac{\hbar\omega}{mv'^2})Q_m(\frac{m(v^2+v'^2)}{4}) \qquad (2)$$

where v and v' are the initial and final electron velocities such that $\frac{mv'^2}{2} - \frac{mv^2}{2} = \hbar\omega$. The total absorption is calculated by integrating over the electron energy distribution

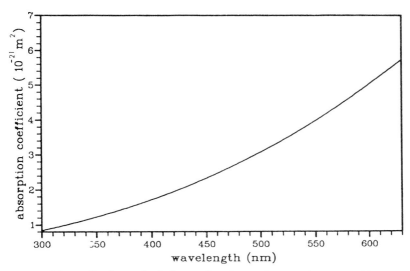

Figure 2: theoretical absorption due to inverse bremsstrahlung

$$\alpha_\omega = \int \eta_\omega F(E) dE \qquad (3)$$

where $F(E)$ is assumed to be a Maxwell-Boltzmann distribution truncated at 20 eV [Cherrington 79, Doughty 83]. The absorption was evaluated from equation 3 by a Runge-Kutta method and the theoretical results are presented in figure 2.

The optogalvanic signal obtained from bremsstrahlung is not easily quantified in terms of a perturbation based model due to the fact that the electron energy is perturbed rather than the density. From the results derived for the electron-atom bremsstrahlung however, it is noted that the total power absorbed by the discharge by this process is of the same order as that of the observed signals and thus the bremsstrahlung can be discounted in this wavelength region unless virtually all of the energy absorbed produces an optogalvanic signal. An approximation for the optogalvanic signal arising can be gained from a fundamental assumption of the Doughty and Lawler model for resonant neon transitions $Z\Delta i + \ell \Delta E = 0$ where ℓ and Z are the positive column length and ballast resistance respectively, and Δi and ΔE are the changes in current and axial electric field. The equation for the axial field is [Brown 59]

$$E = \frac{1.83 \times 10^{-4} x^{1/2} T_e}{\ell} \qquad (4)$$

where x is the fractional energy loss of an electron in collision with a stationary atom. Inverse bremsstrahlung directly perturbs the electric field by exciting electrons to higher energies, thus increasing the electron temperature and causing a change in current. The process of exciting electrons will also produce more electrons with energies in excess of 20 eV, at which point they are strongly absorbed by neon, causing (above 21.5 eV) ionization of the ground state.

Photoionization

Photoionization can be modelled directly by using an expanded version of the Doughty-Lawler [Doughty 83] model. The model includes the metastable $1s_{3,5}$ and resonant $1s_{2,4}$ levels($N_{2...5}$), the electron density(N_e) and the lumped $2p$ levels(U) as their photoionization threshold lies within the studied wavelength region.

The rate equations developed for the discharge are

$$\frac{dN_i}{dt} = P_i N_o N_e + A_{ui} U - W_i M - T N_i \sum_{i=2}^{5} N_i - (S_i + S_i') N_i N_e$$

$$+ N_e \left[\sum_{j=2, j \neq i}^{5} E_{ji} N_j - N_i \sum_{j=2, j \neq i}^{5} E_{ij} \right] + N_o \left[\sum_{j=2, j \neq i}^{5} K_{ji} N_j - N_i \sum_{j=2, j \neq i}^{5} K_{ij} \right]$$

$$i = 2, 3, 4, 5 \quad (5)$$

$$\frac{dN_e}{dt} = \alpha \nu_d N_e + \frac{T}{2} \sum_{i=2}^{5} N_i + N_e \sum_{i=2}^{5} S_i N_i + S_u N_e U - D_a \left(\frac{2.405}{r}\right)^2 N_e \quad (6)$$

$$\frac{dU}{dt} = P_u N_o N_e + N_e \sum_{i=2}^{5} S_i' N_i - S_u N_e U - U \sum_{i=2}^{5} A_{ui} \quad (7)$$

where

- P_i : e^- collisional excitation from the ground state (N_o)
- S_i' : e^- collisional excitation from $N_{2...5} \to U$
- $S_{i,u}$: e^- collisional ionization of $1s_i$, U
- T: metastable collisional ionization
- W_i: diffusion to walls ($N_{3,5}$), trapped radiative decay to N_o ($N_{2,4}$)
- A_{ui}: radiative decay from U to $1s_i$

Also included is the ambipolar diffusion loss of electrons, D_a and single step ionization from the ground state at a rate $\alpha \nu_d$ where α is the first Townsend coefficient and ν_d the electron drift velocity. Collisional mixing of the $1s$ levels is also included with rate constants E_{ij} and K_{ij} for electron and ground state collisions respectively [Smits 79].

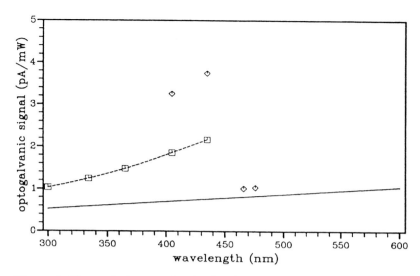

Figure 3: Theoretical optogalvanic signals:—: case 1, □: case 2, ◇: case 3

Three cases of perturbation are included in the analysis:
1. photoionization of highly excited states causing perturbation of N_e (all wavelengths studied here)
2. photoionization of 2p levels perturbing both U and N_e (wavelengths below 478 nm.)
3. Autoionization of the 2p manifold (wavelengths between 390-480 nm.)

Case 1 must be approximated due to lack of knowledge of the populations of all the excited states and photoionization cross-sections at the appropriate wavelengths. The results shown in figure 3 show the signal attributable to a population of $1 \times 10^9 \text{cm}^{-3}$ with $\sigma = 1 \times 10^{-17} \text{cm}^2$. For case 2 the theoretical cross-sections have been evaluated [Duzy 80] and show a decrease with wavelength to a Cooper minima around 180 nm. where the velocity of the ejected electron approaches zero.

Comparing the theory with experiment we can conclude that autoionization is present in the 405, 436 nm. lines and also in the 465-476 nm lines. The greater signal at the first two wavelengths may be attributed to the fact that they lie in the autoionization region of 3 of the 10 2p levels ($2p_{7-9}$ and $2p_{2-4}$ respectively) while the latter lines only autoionize the $2p_1$ level, which has a considerably lower population than the other 2p levels. The signals above 480 nm. are taken to indicate that the photoionization cross-sections of the highly excited states decrease with increasing wavelength.

Measurement of Photoionization Cross-Sections

The results obtained are not state-selective and thus cross-sections for photoionization cannot be deduced directly. However a two-photon technique could make this possible. If one pumps a particular 2p level from the metastable states and then photoionizes with another source the change in the observed continuum signal can be attributed to the change in density of the pumped lower level and a photoionization cross-section may be obtained. Numerical integration of the rate equations described above has shown that pumping a $1s_i - 2p_j$ transition results in a steady state increase of around an order of magnitude in the $2p_j$ population. A small modification to the rate equations described above allows this situation to be modelled and suggests that pumping should cause a 10-15 % increase the continuum signal. Unfortunately this compares with the uncertainty in the signals due to their inherently small size and the low power output of the Hg lamp (4-6 mW) in the blue and near-uv.

Summary

Continuum optogalvanic signals have been observed in neon positive columns and can be attributed predominantly to photoionization of excited states. Inverse bremsstrahlung has been shown to be too small in the studied wavelength region to cause the signals observed. It is noted, however, that at longer wavelengths the process of inverse bremsstrahlung will become much stronger due to its $\simeq \lambda^3$ wavelength dependence and observation of the inverse bremsstrahlung may be employed as a plasma diagnostic to study the electron energy distribution. A theory has been developed to model photo- and auto- ionization of the 2p levels and can be used also to approximate signals arising from higher lying states. Theoretical values for photo- and auto- ionization suggest that these processes are the dominant causes of the continuum signals observed. An experiment has been devised to enable measurements of specific photoionization cross-sections from excited states but the results have so far been inconclusive due to the small size of the signals.

References

[Bekefi 66]   G. Bekefi; *Radiation Processes in Plasmas* Wiley NY (1966)

[Brown 59]   S.C.Brown *Basic Data of Plasma Physics* Wiley NY (1959)

[Cherrington 79] B.E.Cherrington; *IEEE Trans. Elect. Dev ED-26* (2) p.148 (1979)

[Doughty 83]   D.K. Doughty & J.E. Lawler; *Phys. Rev. A 28 (2)* p.773 (1983)

[Duzy 80]   C.Duzy & H.A.Hyman; *Phys. Rev. A* 22 p.1878 (1980)

[Holstein 65]   T. Holstein; *Scientific Paper 65-1E2-Gases-P2* Westinghouse Research Labs, Penn, USA, (1965) (quoted in Bekefi (qv))

[Smits 79]   R.M.Smits & M.Prins; *Physica* 96C p.262 (1979)

[Stark 86]   M.S.Stark, *D.Phil Thesis*, University of York (1986)

Collisional mixing model for the optogalvanic effect on the 6402A ($1s_5$–$2p_9$) line in the neon positive column

R S Stewart, K I Hamad and K W McKnight*

Department of Physics and Applied Physics, the University of Strathclyde, John Anderson Building, 107 Rottenrow, Glasgow G4 ONG, UK.

ABSTRACT

The $2p_9$ level of neon has the special feature that it has only a single radiative decay route to the 1s levels, namely the 6402A ($2p_9$ – $1s_5$) transition. Therefore theoretical modelling of the observed optogalvanic effects on this line should be a useful means of studying collisional mixing. We have developed an optogalvanic model which includes ground–state atom and electron collisional mixing of both the 1s and 2p states.

Preliminary testing of this model has been carried out for optogalvanic measurements in the neon positive column at pressures of 1.5 to 10 Torr and currents of 1 to 10mA. The difficulties of modelling the rather complex 1s mixing and excitation processes are discussed.

INTRODUCTION

In recent years there have been many new applications of optogalvanic spectroscopy (Telle 1990, Sasso 1990). There has also been considerable improvement in quantitative modelling of the optogalvanic effect (Doughty and Lawler 1983, Sasso et al. 1988, de Hoog et al. 1990, Stewart et al. 1990). However, due to the many and complex processes in discharges, it has become clear that theoretical models must be specially designed for each particular case. Nevertheless, for the same reason, there is much to be gained by improved optogalvanic modelling. In the future, the many laser spectroscopic

[* Present address: Rex, Thompson & Partners, Newnhams, Farnham, Surrey GU9 TEQ, UK]

© 1991 IOP Publishing Ltd

techniques now available (optogalvanic, Rydberg-state optogalvanic (Garscadden 1986, Lawler et al. 1986, den Hartog et al. 1988), laser-induced fluorescence, concentration-modulated (Jones 1987), high-resolution optogalvanic (Sasso 1990)) should complement each other to give more complete quantitative spatial and temporal discharge information than has ever before been achieved.

We have worked with neon because it is useful for testing theoretical models due to the appreciable body of the necessary atomic data available in the literature.

For optogalvanic modelling the inter-relation of the radiative and collisional processes is of central importance. We have shown (Stewart et al. 1990) that collisional mixing of the 1s states plays a very important role in determining the magnitude and sign of the optogalvanic effect (OGE) on the neon red lines. The work described in this paper is a theoretical and experimental analysis of the OGE on the 6402A ($2p_9$–$1s_5$) line, which is the single radiative decay route from the $2p_9$ state to the 1s levels. If laser radiation, resonant with this transition, were absorbed by neon atoms under conditions where there were no collisional mixing then the excited atoms would quickly decay from the $2p_9$ level by fluorescence at 6402A giving zero OGE (if the small contribution from ionization of the $2p_9$ upper level is ignored). Generally it is found that this transition gives an easily measurable OGE (although an order of magnitude smaller than for the 6096A and 5945A lines), under discharge conditions found in typical positive column plasmas, so it is clear that collisional mixing is important.

Very little consideration has been given to this line in the past. However, one or two important contributions are worthy of note.

Zalewski et al. (1979), using a hollow-cathode discharge, recorded optogalvanic (OG) signals for this line, which were some eighty times smaller than those for other lines originating on the $1s_5$ state. This was for a current of 15mA in neon at a pressure of 7 torr, conditions under which one would expect considerable collisional mixing.

Fujimoto et al (1983) recognised the value of this line as a clear indicator of the atomic kinetic processes. They reported an interesting analysis of the

6402A and 6334A ($1s_5$–$2p_8$) lines in a positive column plasma to confirm that the population perturbation of the 1s lower level was more important for the OG signal than the population perturbation of the upper 2p state. While the measured $2p_9$ and $2p_8$ population perturbations were about equal, the OG signal magnitudes for these two lines differed by about a factor of 10. Also, the OG signal followed the relaxation time constant ($\simeq 20\mu s$) of the $1s_5$ state rather than those ($\simeq 20$ns) of the $2p_8$ or $2p_9$ states. Fujimoto et al. recognised that the magnitude of the 6402A line signal is determined by mixing. However, the only mixing they refer to is the indirect 1s mixing as a result of 2p mixing from $2p_9$ to the neighbouring $2p_8$ and $2p_{10}$ states. They used a helium–neon gas mixture so a considerable amount of the 2p mixing may have been due to the light mobile helium ground–state atoms.

For completeness and generality of our model we have considered the effect of ionization from the 2p states although this process makes a negligible contribution to the CW laser OGE on lines originating on the neon 1s states. Temporal measurements of the electron density perturbation induced by Uetani and Fujimoto (1984) with a 5ns laser pulse, show that 2p ionization produces only a small positive density perturbation lasting for a few μs whereas the 1s ionization gives rise to a very much larger negative density perturbation which lasts for many tens of μs. For our CW laser experiments we used a low chopping frequency of 93Hz (to ensure that our steady–state rate equations were applicable) corresponding to square excitation pulses lasting about 5ms. The CW laser case may be visualised by integrating the temporal electron density signals of Uetani and Fujimoto over the long 5ms excitation pulse. This shows clearly that the 2p ionization contribution to the OG signal is many times smaller than that of ionization from the 1s states.

EXPERIMENTAL

The experimental set–up, discharge conditions and data acquisition were the same as described previously (Stewart et al. 1990). The state mixing appears to be sensitive to variations in the discharge noise and oscillations so care was taken to optimise the discharge conditions. Variation in the scatter of the OG signal data is an indication of the difference in discharge stability.

THEORY

We will describe the modifications we have made to our theoretical model (Stewart et al. 1990) to improve the method of inclusion of indirect electron collisional coupling of the 1s states via excitation to and redistribution from the 2p states. For clarity of the discussion we recall one of the 1s atom balance equations from our previous paper (1990), the steady-state $1s_5$ rate equation:

$$\begin{aligned}
\dot{M}_5 =\ & n\, S_{g5} N_g - M_5 W_5 - M_5 S_5 n - M_5 T\, (R_2 + M_3 + R_4 + M_5) \\
& - M_5 n S_5' (1-\beta_5') + n\beta_5' (R_4 S_4' + M_3 S_3' + R_2 S_2') \\
& + n(R_2 E_{25} + M_3 E_{35} + R_4 E_{45}) - n M_5 (E_{52} + E_{53} + E_{54}) \\
& + N_g (R_2 K_{25} + M_3 K_{35} + R_4 K_{45}) - N_g M_5 (K_{52} + K_{53} + K_{54}) = 0
\end{aligned} \qquad (1)$$

where the symbols are as defined previously.

The basic perturbed rate equation model is as before with the following modifications:

(A) The bulk branching ratios B_i', describing electron collisional excitation from 1s to 2p followed by radiative decay to the 1s states, are replaced by appropriate terms for every specific 1s-2p excitation process involving this indirect 1s mixing.

(B) Direct collisional excitation from the ground state ($^1S_0$) to the 2p states followed by decay to 1s has been included in a similar way. This was not previously included explicitly.

(C) Direct 1s excitation from the $^1S_0$ ground state has been included explicitly, rather than eliminated from the simultaneous rate equations as was done previously (Stewart et al., 1990).

(D) The X-fractions, defined in our previous model (1990), have been modified to include the new 2p mixing terms.

(E) Ionization from the 2p upper state of the laser perturbed transition is considered for completeness.

(A) The various 1s couplings

Again we will discuss one particular example to illustrate the processes and methodology for their description. We will consider the processes coupling $1s_2$ with $1s_5$.

A(i)   direct electron coupling

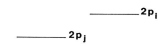

Fig.1a (E_{25},e)

The total rate of $1s_2$–$1s_5$ coupling is given by

$$nR_2 \acute{E}_{25}$$

$$= n R_2 E_{25,e}$$

(already included explicitly in equation (1))

A(ii)   indirect coupling

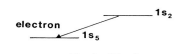

Fig.1b (E_{25},e-r)

$$+ \sum_i \left[nR_2 S_{2i}\right] \left[\frac{A'_{i5}}{D_i}\right]$$

| electron collisional excitation from $1s_2$ to all the 2p states | branching ratios for radiative decay from all 2p to $1s_5$ |

(previously included in equation (1) as part of the β' terms)

A(iii)   indirect coupling

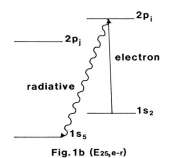

Fig.1c (E_{25},e-e-r)

$$+ \sum_{i,j} \left[nR_2 S_{2i}\right] \left[\frac{nL_{ij}}{D_i}\right] \left[\frac{A'_{i5}}{D_j}\right]$$

electron 2p mixing

(previously included in equation (1) as part of the β' terms)

94   Optogalvanic Spectroscopy

A(iv) indirect coupling

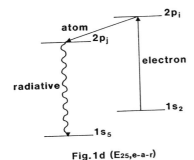

$$+ \sum_{i,j} \left[nR_2 S_{2i} \right] \left[\frac{N_g M_{ij}}{D_i} \right] \left[\frac{A'_{i5}}{D_j} \right]$$

atom 2p mixing

(previously included in equation (1) as part of the β' terms)

Fig. 1d ($E_{25,\text{e-a-r}}$)

Definitions of terms:

S_{ki} :   rate coefficients for electron collisional excitation from $1s_k$ to the 2p states.

$\bar{A}_{ik}$:   radiatively trapped Einstein A-coefficients for decay from $2p_i$ to $1s_k$.

D_i :   total depletion frequencies of the 2p states.

$$D_i = \sum_k \bar{A}'_{ik} + nS_i + \sum_j nL_{ij} + \sum_k N_g M_{ij}$$

where S_i are the 2p ionization rate coefficients.

L_{ij} :   collisional coefficients for electron coupling of the 2p states.

M_{ij} :   collisional coefficients for atom coupling of the 2p states.

All four processes of (A) may be included in the rate equations as a single effective 1s coupling coefficient:

$$E'_{25} = \underset{(i)}{E_{25,\text{e}}} + \underset{(ii)}{E_{25,\text{e-r}}} + \underset{(iii)}{E_{25,\text{e-e-r}}} + \underset{(iv)}{E_{25,\text{e-a-r}}}$$

The corresponding rate terms will then show the following dependencies: (i) i (ii) i (iii) i^2 (iv) i x p, where i is the discharge current (proportional to the electron density) and p is the gas pressure (proportional to the ground-state atom density).

Similar expressions are included in our model for all the other combinations of 1s-state coupling.

(B) Indirect population of the 1s states by electron collision from the ground state ($^1S_0$) to 2p followed by radiative decay to 1s

$$\sum_{i,j} (N_g S_{gi}) \left[\frac{A'_{i5}}{D_i} + \left[\frac{nL_{ij}}{D_i}\right]\left[\frac{A'_{j5}}{D_j}\right] + \left[\frac{N_g M_{ij}}{D_i}\right]\left[\frac{A'_{j5}}{D_j}\right] \right]$$

where S_{gi} are the excitation rate coefficients from $^1S_0$ to 2p.

(C) Direct 1s excitation from the $^1S_0$ ground state

$$n \, N_g \, S_{gk}$$

where S_{gk} are the electron collisional rate coefficients.

(D) The X-fractions

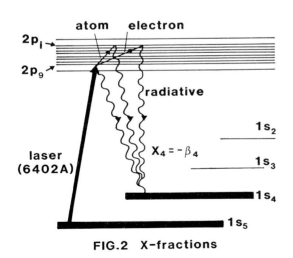

FIG.2 X-fractions

The X-fractions as defined previously (Stewart et al. 1990), give the effective fraction of the absorbed laser power perturbing each of the four 1s states. In the absence of collisional coupling all the X-fractions would be zero for the 6402A transition. The X-fractions have been modified to include 2p mixing. An example is shown in Fig 2. For clarity only the processes involved in X_4 are shown.

$$X_2 = -\left[A'_{92} + \sum_i \left[\frac{nL_{9i} A'_{i2}}{D_i}\right] + \sum_i \left[\frac{N_g M_{9i} A'_{i2}}{D_i}\right]\right] \Big/ D_9$$

The expressions for X_3 and X_4 are similar to that of X_2.

$$X_5 = 1 - \left[A'_{95} + \sum_i \left[\frac{nL_{9i} A'_{i5}}{D_i}\right] + \sum_i \left[\frac{N_g M_{9i} A'_{i5}}{D_i}\right]\right] \Big/ D_9$$

where D_9 is the total depletion frequency (s^{-1}) for the upper level of the 6402A° ($1s_5-2p_9$) transition.

(E) <u>Ionization from the 2p states</u>

$$\sum_{i,j} (nR_2 S_{2i}) \left[\left[\frac{nS_i}{D_i}\right] + \left[\frac{nL_{ij}}{D_i}\right]\left[\frac{nS_j}{D_j}\right] + \left[\frac{N_g M_{ij}}{D_j}\right]\left[\frac{nS_j}{D_j}\right]\right]$$

and similar terms for excitation from $1s_{3,4,5}$.

As a result of the above modifications the balance equations for 1s atoms are as follows (again using $1s_5$ as an example, and ignoring 2p ionization)

$$\begin{aligned}\dot{M}_5 &= nS_{g5}N_g - M_5 W_5 - nS_5 M_5 - M_5 T(R_2 + M_3 + R_4 + M_5) \\ &+ n(R_2 E'_{25} + M_3 E'_{35} + R_4 E'_{45}) - nM_5(E'_{52} + E'_{53} + E'_{54}) \\ &+ N_g(R_2 K_{25} + M_3 K_{35} + R_4 K_{45}) - N_g M_5 (K_{52} + K_{53} + K_{54}) = 0\end{aligned} \qquad (2)$$

EVALUATION OF THE RATE COEFFICIENTS

(a) S_{ki} : . Cross-sections for $1s_{2,3,4,5}-2p_i$ excitation have been published by Beterov and Chebotaev (1967). These agree fairly well with those published by Drawin (1967) for excitation from $1s_{3,4,5}$, those of Samson (1977) for excitation from $1s_5$, and with values averaged over the 1s states by Valignat and Leveau (1981). We also obtained fairly good agreement with cross-sections calculated using the formula of Smits and Prins (1979). In the model we used the cross-sections of Beterov and Chebotaev with our measured electron temperatures to calculate the rate coefficients. In fact, the model is not highly sensitive to variation in these coefficients. Table I shows a typical set of values for a pressure of 3.0 Torr.

Table I: $S_{ki} \times 10^7/cm^3s^{-1}$, $S_{gi} \times 10^{14}/cm^3s^{-1}$

	$2p_1$	$2p_2$	$2p_3$	$2p_4$	$2p_5$	$2p_6$	$2p_7$	$2p_8$	$2p_9$	$2p_{10}$
$1s_2$	7.6	1.9	0	2.5	2.3	7.0	0	1.5	0	0
$1s_3$	0	2.3	0	0	4.6	0	2.6	0	0	0
$1s_4$	0	0.55	0	1.5	0.55	1.5	1.4	3.9	0	2.8
$1s_5$	0	0.33	0	1.0	0.62	2.0	0.48	1.4	5.3	1.8
$g \equiv {}^1S_0$	8.0	3.2	1.5	5.0	2.6	4.2	1.9	1.6	6.1	2.1

(b) The A' coefficients were as calculated in our previous paper (1990), using the escape factor expression given by Fujimoto et al (1982).

(c) Using the total 2p depletion frequencies D_i in the branching ratios, A'_{ik}/D_i for example, conveniently eliminates the need to know the 2p populations.

(d) The 2p electron collisional coupling coefficients L_{ij} are not well known. Smits and Prins (1975) measured the coefficients L_{24} and L_{42} at 300K as 1.6 and $2.6 \times 10^{-4} cm^3 s^{-1}$ respectively. We have used these values as a guide for choosing values of the other 2p coupling coefficients. We assumed that the coupling to the nearest neighbouring 2p states was strongest and decreased rapidly for energetically more distant states (See Table II).

Table II: Assumed relative values of electron 2p mixing rate coefficients L_{ij}

j	1	2	3	4	5	6	7	8	9	10
i	$2p_1$	$2p_2$	$2p_3$	$2p_4$	$2p_5$	$2p_6$	$2p_7$	$2p_8$	$2p_9$	$2p_{10}$
1 $2p_1$	–	0.1	0.08	0.06	0.04	0.02	0.01	0.008	0.006	0.004
2 $2p_2$	0.1	–	1	0.6	0.3	0.15	0.07	0.035	0.02	0.006
3 $2p_3$	0.08	1	–	1	0.6	0.3	0.15	0.07	0.04	0.008
4 $2p_4$	0.06	0.6	1	–	1	0.6	0.3	0.15	0.07	0.01
5 $2p_5$	0.04	0.3	0.6	1	–	1	0.6	0.3	0.15	0.02
6 $2p_6$	0.02	0.15	0.3	0.6	1	–	1	0.6	0.3	0.04
7 $2p_7$	0.01	0.07	0.15	0.3	0.6	1	–	1	0.6	0.06
8 $2p_8$	0.008	0.04	0.07	0.15	0.3	0.6	1	–	1	0.08
9 $2p_9$	0.006	0.02	0.04	0.07	0.15	0.3	0.6	1	–	0.1
10 $2p_{10}$	0.004	0.006	0.008	0.01	0.02	0.04	0.06	0.08	0.1	–

For simplicity we used a multiplying factor L in our computation to allow us to increase or decrease all of the L_{ij} values, depending on the electron temperature, to fit theory with experiment. These coefficients are of fundamental importance for the 6402A line OGE and will be discussed in detail later.

(e) The 2p ground-state atom collisional coupling coefficients M_{ij} are much better known than the equivalent electron coefficients. Steenhuysen (1981) published a complete set of values, with their temperature dependence, while more limited sets had been published by Coolen et al (1978) and Grandin et al

(1975). These values showed good agreement. We used the Steenhuysen values as a starting point. Again, we used a multiplying factor to scale these coefficients to fit theory with experiment. This 2p mixing by atomic collisions is also of central importance to the OGE on the 6402A line and will be discussed further below.

(f) The rate coefficients S_{gi}, for direct 2p excitation from the ground state, were calculated using the cross-sections of Valignat and Leveau (1981) with our measured electron temperatures. Table I shows values at 3.0 Torr. Again, the model is not highly sensitive to variations in these coefficients.

(g) The rate coefficients S_{gk}, for 1s electron collisional excitation from the ground state, were evaluated from cross-sections published by Phillips *et al* (1985) and are shown in Table III. The values from Valignat and Leveau (1981), for $1s_4$ and $1s_5$, are given in brackets.

Table III: Rate coefficients S_{gk} for excitation of the 1s states
(Units: $10^{-11} cm^3 s^{-1}$)

P/Torr	1.5	2.0	3.0	4.0	10.0
S_{g2}	0.45	0.32	0.25	0.24	0.23
S_{g3}	0.042	0.036	0.030	0.029	0.028
S_{g4}	0.060 (0.51)	0.042 (0.41)	0.033 (0.34)	0.033 (0.33)	0.030 (0.31)
S_{g5}	0.22 (0.77)	0.15 (0.62)	0.12 (0.51)	0.11 (0.49)	0.12 (0.22)

(h) The value of the 2p ionization rate coefficient ($2 \times 10^{-8} cm^3 s^{-1}$) was taken from the publication by Kane (1984). As mentioned above, this makes negligible contribution to the OG signal. Using the 2p populations measured by Valignat and Leveau (1981) for similar positive column conditions, a simple calculation shows that 2p ionization is about three orders of magnitude lower than 1s ionization.

COMPARISON OF THEORY AND EXPERIMENT

The atomic data, such as the 1s populations, discharge impedance, 1s atom wall losses etc. are as discussed previously (Stewart et al 1990). As before, the electron and excited atom densities are averaged over the tube. We again discuss experiment and theory in terms of Δ where

$$\Delta = \left[\frac{\Delta i}{Q}\right]\left[1 + \frac{Z}{Z_d}\right] = \frac{\partial i}{\partial n} \frac{K_Q}{K_n} \qquad (3)$$

Variation of the optogalvanic current perturbation Δi, with current i, is dominated by the dynamic impedance of the discharge, Z_d. We are mainly interested in improving the modelling of the OGE so that it may be employed to obtain quantitative information on the atomic kinetics. Therefore we wish to remove the dominant effect of the discharge impedance term. The form of equation (3) allows direct investigation of the kinetic information which is contained in the term $\partial i/\partial n . K_Q/K_n$. For our discussion we shall refer to the quantity Δ as the OG kinetic term and set

$$\Delta_{theory} = \frac{\partial i}{\partial n} \frac{K_Q}{K_n} \qquad (4)$$

and

$$\Delta_{exp} = \left[\frac{\Delta i}{Q}\right]\left[1 + \frac{Z}{Z_d}\right] \qquad (5)$$

We must point out that there is some danger in doing this because both Δi and $1 + Z/Z_d$ vary strongly with i. Any uncertainties in Δi and the impedance term may swamp the i-dependence of the kinetic term. More will be said about this later.

Our results are presented in Fig 3. For 1.5 Torr and 2.0 Torr for which very quiet discharges could be obtained, the agreement between theory and experiment is excellent. For 3.0 Torr there is some discrepancy between theory and experiment. We believe that this is due to the onset of new discharge processes not included in the model.

We carried out computational fitting to investigate the observed current and pressure behaviour of Δ. We used values of coefficients derived from the

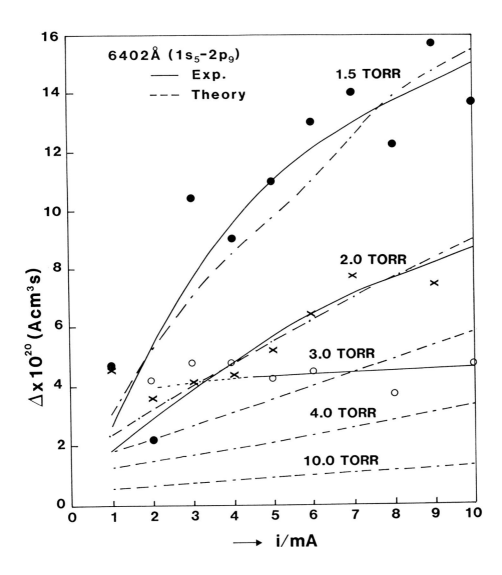

Fig 3 Variation of the optogalvanic kinetic term Δ with current and pressure

literature, except for the three sets involved in the direct and indirect 1s coupling, namely the 1s electron and 2p atom and electron coupling coefficients, $E_{kl,e}$, M_{ij} and L_{ij} respectively. However, even for these, our new optogalvanic model requires values which are very close to those indicated in published literature.

We first fitted the direct 1s coupling coefficients $E_{kl,e}$, for the 6096A ($1s_4-2p_4$) sign-change line which we previously showed (Stewart et al 1990) to be a sensitive indicator of 1s coupling. In this computation we used the M_{ij} values of Steenhuysen (1981) and a scaling factor $L = 2 \times 10^{-4} cm^3 s^{-1}$ with Table II, in agreement with Smits and Prins (1975). Our new model gave the values shown in Table IV.

Table IV: Values of the 1s direct electron collisional coupling coefficients $E_{kl,e}$ (fitted for the 6096A line) (Units: $10^{-7} cm^3 s^{-1}$)

p/Torr	$E_{23,e} = E_{34,e} = E_{35,e}$	$E_{24,e} = E_{25,e}$	$E_{45,e}$
1.5	1.2	3.0	8.0
2.0	1.2	3.0	7.6
3.0	1.2	3.0	7.4

These values are about an order of magnitude lower than those obtained using our previous model (Stewart et al 1990) which did not specify the details of the indirect coupling via the 2p levels. The values of Table IV are now in close agreement with a number of publications (Ionikh et al 1977, Valignat and Leveau 1981, Tachibana et al 1982 and Philosof and Blagoev 1987), although there is considerable variation amongst these.

After determination of the $E_{k\ell,e}$ coefficients, the new theory was fitted to the 6402A line results using the scaling factors M and L for the coefficients M_{ij} and L_{ij}. For 1.5, 2.0 and 3.0 Torr best fits were obtained for $M = 3.0 \times 10^{-12} cm^3 s^{-1}$ (which compares well with the Steenhuysen value of $1.0 \times 10^{-11} cm^3 s^{-1}$ (Steenhuysen 1981)), along with the values of L given in Table V (which must be employed with Table II).

Table V: Values of the 2p scaling factors L and M

p/Torr	Te/10^{4}K	L/10^{-4}cm^{3}s^{-1}	M/10^{-12}cm^{3}s^{-1}
1.5	5.3	1.8	3.0
2.0	5.0	1.0	3.0
3.0	4.6	0.6	3.0
4.0	4.5	~0.4	3.0
10.0	4.4	~0.3	3.0

For example, these results mean that $L_{9,10} = L_{10,9} = 1.8 \times 10^{-5}$cm^{3}s^{-1} for an electron temperature of 5.3×10^4K. We feel that values of about 10^{-4}cm^{3}s^{-1}, for the coupling of the other near neighbours of Table II, are of the right order, because they are much closer in energy than the $2p_9$ and $2p_{10}$ states.

The electron 2p coupling coefficients are strongly dependent on the electron temperature as expected, while the atomic 2p coefficients are independent of Te, as expected for constant gas temperature.

At low currents the 1.5 and 2.0 Torr curves are sensitive to the atomic 2p mixing so the M scaling factor was determined mainly by the low current regime. As the current rises the electron 2p mixing mainly determines the gradient of the curves.

For interpreting our findings it is important to note that the effects of 2p coupling appear in two important parts of our model, firstly in the rate equations and secondly, in the X-fractions.

The 1s mixing appears explicitly in the balance equations and hence in the perturbed rate equations. Three of the coupling terms depend directly on the current i and the other on i^2. It is interesting to note that in terms like $\partial \dot{M}_5/\partial n$, in the perturbed equations, this i^2 electron-electron coupling will have its coefficient effectively doubled because of the differentiation with respect to n. Also, the 2p coupling due to atom collisions means that Δ will also be pressure dependent.

The expressions above show clearly that all the X's will immediately become zero if L_{9j} and M_{9j} are set to zero, so any variation of these away from zero will have a direct effect on Δ.

All of these considerations mean that Δ should show a strong current dependence as well as being pressure dependent.

At low pressures (< 3 Torr) it would appear that atom mixing is small. At low currents even electron mixing is small so that the X values of the 6402A line are small and therefore Δ is small, as evidenced by the fact that the curves appear to start from the origin. Then as i increases greater electron mixing of the 2p levels raises the X's and Δ increases with i, faster at 1.5 Torr because of the higher electron temperature.

Although there is considerable scatter on the data, both the 1.5 and 2.0 Torr cases would seem to indicate that a saturation curve was a reasonable assumption. However, there is an interesting feature appearing in both theoretical curves. There is a slight flattening at about 4-5mA and then the gradient increases again. This may be due to the onset of the influence of the i^2 coupling terms after the i-dependent terms had given the main contribution at low i. The precision of our data does not allow any definite conclusions to be drawn.

It is also worth noting that the magnitude of Δ_{theory} is sensitive to the values of the S_{gk} coefficients for electron collisional excitation of the 1s states. The excellent agreement of theory and experiment for the 6402A line shown in Fig 3, and the approximate agreement of the direct 1s coupling coefficients obtained using the 6096A line with a number of literature sources, gives us confidence in the S_{gk} values of Phillips et al (1985). The values of Valignat and Leveau (1981) may be too high (see Table III).

We propose the following reason for the sensitivity of Δ to variations in the S_{gk} coefficients. A decrease in one of these has just the same kind of effect as the laser perturbation itself, namely a decrease in the 1s population.

These coefficients therefore directly affect the magnitude of Δ_{theory}.

Although we have expressed a warning of the danger involved in writing the experimental equation as in equation (5), we are confident that our new model gives an excellent description of the optogalvanic effect on the 6402A ($1s_5 - 2p_9$) transition in neon positive column discharges at 1.5 Torr and 2.0 Torr. Although the model is very detailed, it indicates very clearly which are the important processes causing the optogalvanic effect on this line.

Fig 3 includes the theoretical predictions of the model for pressures of 4 and 10 Torr, using the extrapolated L and M values of Table V. We believe the L values used are reasonable and we are confident that M remains constant with pressure so its value should be correct. The computational results would appear to indicate that the 6402A line kinetic term is lowered at higher pressures mainly because of the reduced electron 2p coupling due to the reduced electron temperature. At present we are carrying out new experiments to investigate the 6402A line optogalvanic behaviour at higher pressures.

CONCLUSIONS

We have developed and tested a new rate-equation model for the OGE on the 6402A ($1s_5 - 2p_9$) transition in the neon positive column. This line is of special interest for model testing because the existence of an OG signal depends on collisional mixing amongst the 2p states. This work gives further proof that our mathematical formulation, removing the strong influence of the discharge impedance by defining an OG kinetic term, is a useful means of obtaining quantitative information on the discharge kinetic processes (see also Stewart *et al* 1990 for studies of the 6096A and 5945A transitions using this method).

Remarkable agreement between our new theory and experiment has been achieved for the well-behaved discharges at 1.5 and 2.0 Torr. The model confirms many literature coefficients and gives new information on electron collisional energy transfer among the 2p states.

Further investigations are planned to directly investigate the various 1s coupling processes. The results of this work and further testing of the model will be published elsewhere.

ACKNOWLEDGEMENTS

We would like to thank Drs R Illingworth and I S Ruddock for the use of their tunable dye laser system in the early stages of this work. We are also grateful to the SERC and the Embassy of the Republic of Iraq for postgraduate support.

It is a pleasure to recognise the technical help of Mr J Barrie, Mr R Beattie, Mr R W Dawson and Mr D Graham.

REFERENCES

Beterov I M and Chebotaev V P 1967, Opt. Spectroscop. **23** 467.

Coolen F C M, van Shaik N, Smits R M M, Prins M and Steenhuysen L W G 1978 Physica **93C** 131.

Doughty D K and Lawler J E 1983, Phys. Rev. A **28** 773.

de Hoog F J 1990 in these Proceedings

den Hartog E A, Doughty D K and Lawler J E 1988, Phys. Rev. A **38** 2471.

Drawin H W 1967 Rapport EUR-C.E.A.-F.C.-383, Révisé Association Euratom - C.E.A., Fontenay-aux-roses (1966).

Fujimoto T, Goto C and Fukuda K 1982, Phys. Scr. **26** 443.

Fujimoto T, Uetani Y, Sato Y, Goto Y, Goto C and Fukuda K 1983, Opt. Comm. **47(2)** 111.

Garscadden A 1986 in 'Rad^{ve}. Procs. in Discharge Plasmas' (NATO, ASI Series B: Physics) vol. 149 (ed. J M Proud and L H Leussen) (New York: Plenum) pp547-68 and references therein.

Ionikh Yu Z, Penkin N P and Samson A V (1977), Opt. Spectrosc. **43(5)** 491.

Grandin J P, Hennecart D, Husson X, Lecler D, Vienne J F and Barrat-Ramsbosson M 1975, Le Journal de Physique **36** 78.

Jones W J 1987, J. Chem. Soc., Faraday Trans. **83** 693.
Kane D M 1984, J. Appl. Phys. **56** 1267.

Lawler J E, Doughty D K, den Hartog E A and Salih S, 1986 in 'Rad^{ve}. Procs. in Discharge Plasmas' (NATO, ASI Series B: Physics) vol 149 (ed J M Proud and L H Leussen) (New York: Plenum) pp547-68 and references therein.

McKnight K W and Stewart R S (1990) in these Proceedings.

Phillips M H, Anderson L W and Lin C C 1985, Phys. Rev. A **32(4)** 2117.

Pilosof N and Blagoev 1988, J. Phys. B: At. Mol. Opt. Phys **21** 639.

Samson A V 1977, Opt. Spectrosc. **42(3)** 321.

Sasso A, Ciocca M and Arimondo E 1988, J. Opt. Soc. Am. **85** 1484.

Sasso A in these Proceedings and references therein.

Smits R M M and Prins M 1975, Physica **80C** 571.

Steenhuysen L W G 1981
    Beitr. Plasmaphys. **21** 301.

Stewart R S, McKnight K W and Hamad K I 1990, J. Phys. D. Appl. Phys. **23** 832.

Tachibana K, Harima H and Urano Y 1982, Jap. J. Appl. Phys. **21(11)** 1529.

Telle H H 1990 in these Proceedings and references therein.

Uetani Y and Fujimoto T 1984, Opt. Comm. **49(4)** 258.

Valignat S and Leveau J 1981, Physica **104C** 441.

Zalewski E F, Keller R A and Engelman R 1979, J. Chem. Phys. **70(02)** 1015.

Calculation of resonance radiation trapping

F Vermeersch and W Wieme

Laboratorium voor Natuurkunde, Faculteit Toegepaste Wetenschappen, Rijksuniversiteit Gent, Rozier 44, 9000 Gent, Belgium.

ABSTRACT: An overview is given of the different methods that were developed for calculating the trapping of resonance radiation. Analytical as well as Monte Carlo techniques are included. Theories assuming partial frequency redistribution are considered and comparison is made with recent experiments performed in our laboratory.

1. INTRODUCTION.

The trapping of resonance radiation plays a major role in most laboratory and technical gas discharges. Therefore a theoretical and experimental investigation of this phenomenon is important. A good understanding of the trapping of resonance radiation leads to a better development of lasers and lamps (Anderson 1985) and more specifically, it also leads to a better interpretation of optogalvanic signals (Stewart 1990).

The trapping of resonance radiation can be described by considering a gas of two level atoms. When the gas is excited, by means of a gas discharge or by means of laser excitation, a number of atoms will be brought into the upper level. These excited atoms will decay to the ground level by emission of a photon with a frequency ν centered around ν_0 with $\nu_0 = \Delta E/h$. The decay rate of these excited atoms is equal to A_{ul}, with A_{ul} the Einstein coefficient for spontaneous emission. The excited atom will remain an average time $\tau_n (=1/A_{ul})$ in the upper level before decaying to the ground state. The emitted photon will then travel through the gas and can be reabsorbed by an atom at another position in the gas. If the ground state density is high the probability of reabsorption will be large. The newly excited atom can, again after an average time τ_n, re-emit a photon.

© 1991 IOP Publishing Ltd

This process of emission and reabsorption goes on until the photon eventually reaches the boundary of the gas. A resonance photon will make many stops before it escapes from the gas and will in that way prolong the lifetime of the excitation in the gas. The excited state will spend a time $g\tau_n$ in the cell with g the number of stops the photon makes before escaping. We can picture the movement of the photon in the gas as a kind of "random walk" (see Figure 1).

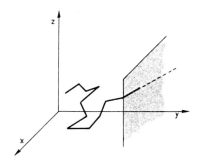

Fig. 1. "Random walk" of the photon in the gas.

In the calculation of the resonance radiation trapping we want to determine how fast the excitation leaves the experimental cell. In the following, a review is given of different theories, developed for this purpose. Comments are made on their applicability in different experimental situations.

2. THE MODIFIED DIFFUSION THEORY.

A first attempt to describe the trapping of the resonance states was made by Compton (1923). His theory was based on the diffusion equation and gave poor agreement with experiments at very low opacities. Milne (1926) was the first to introduce the concept of emission and absorption of the resonance photons to describe the measured trapping of the resonance light. He derived the following modified diffusion equation:

$$\nabla^2 \left[N_u(\vec{r},t) + \tau_n \frac{\partial}{\partial t} N_u(\vec{r},t) \right] = 4k^2 \tau_n \frac{\partial}{\partial t} N_u(\vec{r},t) \qquad (1)$$

$N_u(\vec{r},t)$ is the upper state density at position $\vec{r}$ and time t, τ_n is the natural lifetime of the excited state and k is a frequency independent absorption coefficient given in m^{-1}. The form of this equation is that of a diffusion equation, except for the second term on the left hand side. This term describes the natural decay of the excited states. This term was omitted in Compton's theory giving poor agreement with experiment. It is worth noting that the absorption coefficient k, on the right hand side in the equation, is taken to be the same for all photon frequencies. This means that all photons have the same mean free path. Milne's theory was improved by Blickensderfer (1976) who included effects of line shape into the absorption coefficient by defining an equivalent opacity $\bar{k}L$ using:

$$e^{-\bar{k}L} = \frac{\int_0^{+\infty} e^{-k(\nu)L} \phi(\nu) d\nu}{\int_0^{+\infty} \phi(\nu) d\nu} \qquad (2)$$

where $\phi(\nu)$ is the line shape, which can be the Doppler-, the Lorentz- or the Voigt line shape. $k(\nu)$ is the absorption coefficient at frequency ν and L is the typical distance in the trapping problem. (e.g. the typical distance in an infinite cylindrical geometry is equal to the radius of the cylinder). In many experimental configurations a cylindrical geometry is used to confine the gas. The general solution of Milne's modified diffusion equation in this geometry is given by the following expansion:

$$N_u(r,t) = \sum_{m=1}^{\infty} A_m J_0(\lambda_m r) e^{-(t/\tau_m)} \qquad (3)$$

where J_0 is the Bessel function of zeroth order.
The boundary condition that no inward flux is present at the wall of the experimental cell, restricts λ_m's to values that must satisfy the following equation:

$$\lambda_m R J_1(\lambda_m R) - 2kJ_0(\lambda_m R) = 0 \qquad (4)$$

with R the radius of the cylinder.
The decay times τ_m in the expansion are given by:

$$\tau_m = \tau_n \left[1 + 4\left(\frac{k}{\lambda_m}\right)^2 \right] \qquad (5)$$

At "early times" all the terms in the expansion will contribute to the behavior of the upper state density $N_u(r,t)$. At "late times" the upper state density is described by the term with the lowest decay rate. In consequence the decay of the resonance radiation intensity "at late times" is given by a simple exponential decay, with decay time τ_1. It is this lowest decay rate that is mostly measured in the experimental investigation of the resonance radiation trapping.

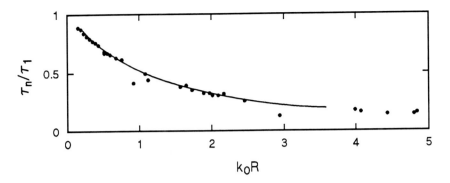

Fig. 2. Measurements by Romberg(1988) on the Li 670.67 nm radiation compared with Milne's theory (●,measurements;———,theory).

From Figure 2 it is seen that the modified diffusion theory is very successful in describing the experimental data at low opacities. The data are taken from experiments by Romberg (1988) on the trapping of the 670.67 nm resonance radiation in Li. The agreement is very good for opacities up to 4. At higher opacities the theory fails due to the assumption of the same mean free path for all photon frequencies. This assumption is only valid if the absorption coefficient varies little over the frequency spectrum of the photons. However, the absorption coefficient is sharply peaked around the central frequency v_0 and drops off rapidly for higher and lower frequency values. At higher opacities the transport of the photons will be mainly in the wings of the spectral profile, where the opacity is smaller, i.e. where the free path of the photon is larger. Holstein was the first to make this criticism of Milne's theory.

3. THE HOLSTEIN-BIBERMAN THEORY.

Around the same time Holstein (1947) and Biberman (1947) developed a radiation diffusion theory, based on escape probabilities of photons. To develop his theory, Holstein made the following assumptions: (1) no external radiation field acts on the medium. (2) the emission and absorption profile are the same $\phi(v) \equiv \psi(v)$, i.e. complete frequency redistribution (CFR), no correlation exists between the frequency of the absorbed and emitted photon. (3) the emission after absorption is isotropic. This assumption is valid in cases of strong imprisonment where the absorbed emission comes nearly from all directions so that the emission is approximately isotropic. (4) the ground level density is constant throughout the medium. This is valid, when the excitation is so low that the de-excitation to the lower level does not alter the ground state density. In his theory Holstein derives what he calls the transmission factor T(ℓ), defined as the probability that *a photon* will travel a distance ℓ through the gas without being absorbed. The probability that a photon *with frequency* ν will travel a distance ℓ without being absorbed is given as $e^{-k(\nu)\ell}$, here k(ν) represents the absorption coefficient at frequency ν expressed in m^{-1}. From this expression it is clear that the mean free path of the photon is dependent on the photon frequency. By averaging over the probability that a photon is emitted in the interval [ν,ν+dν] Holstein derived the transmission factor T(ℓ):

$$T(\ell) = \int_0^{+\infty} \phi(\nu) e^{-k(\nu)\ell} d\nu \qquad (6)$$

For high $k(\nu_0)\ell$ values ($k(\nu_0)\ell \gg 1$), Holstein derived asymptotic approximations for two types of line broadening, namely Doppler- and Lorentz broadening:

$$\text{Doppler:} \qquad T(\ell) = \frac{1}{k(\nu_0)\ell \left[\ln\left(k(\nu_0)\ell\right) \right]^{1/2}} \qquad (7)$$

$$\text{Lorentz:} \qquad T(\ell) = \frac{1}{\left[\pi k(\nu_0)\ell\right]^{1/2}} \qquad (8)$$

When the collision rate is low the emission and absorption line will be characterized by the Doppler profile. At high collision rates, where effects of pressure broadening gain more importance, the absorption-emission profile will be characterized by the Lorentz line.

It is interesting to look at the probability that a photon will leave the experimental cell without being absorbed. This probability is called the escape factor. An in depth study on this subject is given in a series of articles by Irons (1979a,b,c).

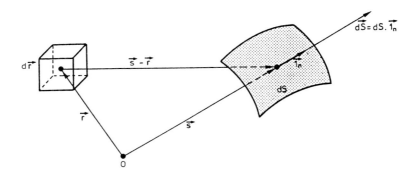

Fig. 3. Calculation of the escape factor.

With the use of the above mentioned transmission factor Irons (1979a) derived the escape factor at a position $\vec{r}$ in the medium:

$$\theta(\vec{r}) = \frac{1}{4\pi} \int_S \frac{1}{|\vec{s}-\vec{r}|^2} \frac{(\vec{s}-\vec{r})}{|\vec{s}-\vec{r}|} \, d\vec{S} \, T\left(|\vec{s}-\vec{r}|\right) \tag{9}$$

The integration is taken over the whole outer surface of the medium (see Figure 3). An alternative way of defining $\theta(\vec{r})$ is:

$$\theta(\vec{r}) = 1 - \int_V G(\vec{r},\vec{r}') \, dV' \tag{10}$$

$G(\vec{r},\vec{r}')$ is the probability that a photon emitted at $\vec{r}$ is absorbed in a unit volume around $\vec{r}'$ in the gas. The integration is taken over the whole volume of the gas. This function is related to the transmission factor through:

$$G(\vec{r},\vec{r}')\Delta r = -\frac{1}{4\pi\ell^2} \frac{\partial}{\partial \ell} T(\ell) \quad \text{with } \ell = |\vec{r}-\vec{r}'| \tag{11}$$

By averaging $\theta(\vec{r})$ over the spatial distribution of the excited states, the probability θ, that a photon will leave the experimental cell without being absorbed is obtained as:

$$\theta = \int_V p(\vec{r})\theta(\vec{r})d\vec{r} \quad (12)$$

where $p(\vec{r})$ is the normalized spatial distribution of the excited states. From equations (7-12) it is obvious that the escape factor (12) is dependent on: (1) the spatial distribution of the excited states, (2) the emission-absorption line shape and (3) the opacity of the medium. It is possible to calculate the upper and the lower limit for the escape factor in simple geometries like a cylinder or a slab. The upper limit can be derived by using a uniform distribution of the excited states in the cell. The lower limit is obtained by use of a δ-distribution, where the excitation is on the axis of the cylinder. In Figures 4a,b the upper and lower limit, calculated for a cylindrical geometry are shown, this respectively for Doppler- and Lorentz broadening. Notice that the escape factor is more sensitive to the spatial distribution in the case of Doppler broadening than in the case of Lorentz broadening. In the case of the Lorentz profile the transport will be mainly in the optically thin wings of the line, resulting in a weak dependence on the spatial distribution.

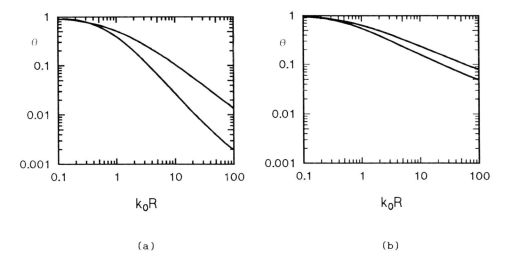

(a) (b)

Fig. 4. Upper and lower limit for the escape factor in a cylindrical geometry for (a) Doppler broadening and (b) Lorentz broadening.

In many laboratory experiments pulsed sources are used to study resonance radiation trapping. In these sources the excitation is turned off at t=0 and the decay of the population density is measured. Holstein (1947) wrote down the following equation to describe the upper state density in such a source:

$$\frac{\partial}{\partial t} N_u(\vec{r},t) = -A_{ul} N_u(\vec{r},t) + A_{ul} \int_{V'} N_u(\vec{r},t) G(\vec{r},\vec{r}\,') dV' \qquad (13)$$

The first term on the right hand side represents the natural decay of the excited states. The second term describes the reabsorption at $\vec{r}$ of photons emitted at $\vec{r}\,'$. The general solution of this integrodifferential equation is given by:

$$N_u(\vec{r},t) = \sum_{j=1}^{\infty} c_j n_j(\vec{r}) e^{-g_j A_{ul} t} \qquad (14)$$

In this expansion the coefficients c_j are determined by the initial spatial distribution. The spatial functions $n_j(\vec{r})$ and the parameters g_j are geometry dependent. The behavior at "late times" will be described by the term in the expansion with the lowest g_j value. The spatial function $n_1(\vec{r})$, corresponding to this lowest value, is currently called the fundamental mode. The decay of the upper state density at "late times" will be a simple exponential decay at every spatial position in the cell and is given by:

$$N_u(\vec{r},t) = c_1 n_1(\vec{r}) e^{-\beta t} \quad \text{with} \quad \beta = g_1 \frac{1}{\tau_n} \qquad (15)$$

β is called the decay rate at "late times". The parameters g_j are determined by substituting the equation (14) into the transport equation. The solution of the equation then becomes an eigenvalue problem. Holstein used the variational Ritz method to determine a solution for the lowest eigenvalue g_1 in a cylindrical geometry of radius R. As part of the Ritz method the $n_1(r)$ function was approximated by a linear combination of a zeroth, second and fourth order function:

$$n_1(r) = \sum_{j=1}^{m} a_j p_j(r) \qquad (16)$$

The functions $p_j(r)$ are called trial functions (a discussion about the choice and validity of these trial functions can be found in an article by Preobrazhenski (1968)). The values of a_j are determined by minimizing g_1 in the eigenvalue equation. Holstein arrives at the following results in the cases of Doppler- and Lorentz broadening:

Doppler: $$g_1 = \frac{A}{k(\nu_0)R\left[\ln\left(k(\nu_0)R\right)\right]^{1/2}} \qquad A=1.60 \qquad (17)$$

Lorentz: $$g_1 = \frac{B}{\left[\pi k(\nu_0)R\right]^{1/2}} \qquad B=1.125 \qquad (18)$$

From these results and equation (15) it is clear that the decay rate at "late times" depends upon the optical thickness $k(\nu_0)R$ of the medium and the line shape which is used. Irons (1979b) shows that g_1 is nothing else than the escape factor associated with the spatial distribution $n_1(r)$. Solutions of the Holstein-Biberman equation were also derived by other authors, who arrived at comparable results. Values for the A and B parameters are given in the following table.

author	A (Doppler)	B (Lorentz)	method
Holstein	1.60	1.125	variational
Preobrazhenski	-	1.1247	calculation
Payne	1.575	1.123	analytical
Van Trigt	1.57560	1.12274	analytical
Klots	1.60	-	Monte-Carlo
Vanmarcke	1.8	1.13	calculation

In Figure 5 the decay rates at "late times", calculated with Holstein's theory, are shown for the 146.96 nm resonance radiation in Xe . The calculations are performed for a cylindrical geometry of 1.9 cm radius in the case of Doppler- and natural+pressure broadening respectively. On this double logarhitmic scale the approximate p^{-1} behavior of the decay rate in the case of Doppler broadening is clearly seen (p is the gas pressure at 300 K). The constancy of the decay rate at higher pressures (around 10 Torr) is typical for the natural+pressure broadening. Holstein's theory is limited to the lowest eigenmode or the fundamental mode. Payne (1970) and Van Trigt (1969) completed the theory by considering the higher modes. Because the fundamental mode is positive for all r the higher modes must be negative for some r-values. These higher modes are not physical in the sense that the system cannot relax to one of these modes. However by making use of them in formula (14) the behavior of the system can be described at "early times" .

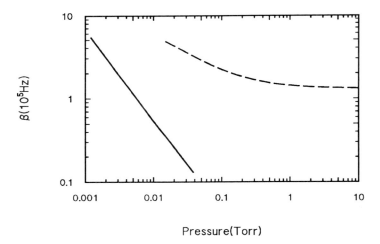

Fig. 5. Decay rates at "late time" for the 146.96 nm radiation in Xe in a cylindrical geometry of 1.8 cm radius τ_n=4.6 ns.(——,Doppler broadening;-----,Lorentz broadening.)

In Figure 6a the recent experiments by Romberg (1988) in Lithium performed at low pressure, are shown. The center solid line plotted for $k_0 l < 4$ is calculated with Milne's theory. The center solid line for $k_0 l > 4$ shows the results by Holstein. For comparison the upper and the lower limit of the escape factor are included in the Figure. We note a good agreement between the Holstein-Biberman theory and the experiment.

However notice that the calculation by Holstein fails at low opacities. This is due to the approximation he used for the $G(\vec{r},\vec{r}')$-function which was not correct at low opacities. However Phelps (1958) showed that good agreement can be found with experimental data when a more correct $G(\vec{r},\vec{r}')$-function is used to solve the Holstein-Biberman equation. Due to the low collision rate at these low pressures the broadening is described by a Doppler line. In Figure 6b we compare experiment and theory at higher pressures where, due to the high collision rate, the emission absorption profile is described by the Lorentz line shape. The data are taken from Wieme et al. (1979). From this figure it is clear that the Holstein-Biberman theory describes the decay rate for the Xe 146.96 nm resonance radiation very well for pressures above 2 Torr.

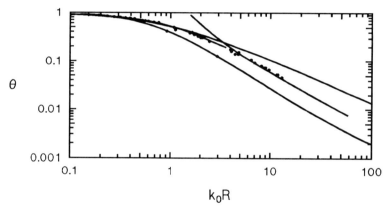

Fig. 6a. Comparison of experiment with the Holstein theory in the case of Doppler broadening.

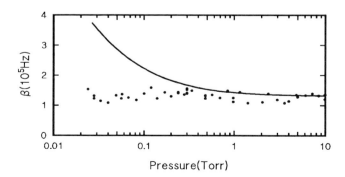

Fig. 6b. Comparison of experiment with the Holstein theory in the case of natural+pressure broadening (●, experiment;——— Holstein's theory).

4. PARTIAL FREQUENCY REDISTRIBUTION.

One of the basic assumptions of Holstein's theory is that of complete frequency redistribution, i.e. no correlation exists between the frequency of the absorbed and emitted photons. Payne (1974) showed that correlation between absorbed and emitted photon frequencies can have a noticeable effect on the trapping of resonance radiation. He therefore wrote down the following transport equation:

$$\frac{\partial}{\partial t} N_u(\vec{\rho},x,t) = - A_{ul} N_u(\vec{\rho},x,t) + S(\vec{\rho},x,t) - AN_u(\vec{\rho},x,t)$$

$$+ A_{ul} k_0 R \pi^{1/2} \int_{-\infty}^{+\infty} dx' \int_V d\vec{\rho}\,' N_u(\vec{\rho}\,',x,t) \bar{R}(x',x) \frac{e^{-k(x')R|\vec{\rho}\,'-\vec{\rho}|}}{4\pi|\vec{\rho}\,'-\vec{\rho}|^2} \quad (19)$$

For the definition of the different terms and symbols we refer to Post (1986a). The partial frequency redistribution (PRD) is taken into account by the angle averaged redistribution function $\bar{R}(x',x)$. This function represents the probability that a photon of frequency x will be emitted when a photon of frequency x' is absorbed. The frequencies are expressed in Doppler units $x=(\nu-\nu_0)\bar{v}/\nu_0 c$ where $\bar{v}$ is the average velocity of the atom.

Before turning our attention to the solution of the transport equation, we will first elaborate on the physical mechanism of partial frequency redistribution. Let us therefore define, following Hummer (1962), the global redistribution function $R(\nu',\vec{n}\,',\nu,\vec{n})$ such that $R(\nu',\vec{n}\,',\nu,\vec{n})d\nu'd\Omega'd\nu d\Omega$ represents the probability that a photon of frequency ν from $[\nu,\nu+d\nu]$ will be emitted in the solid angle $d\Omega$ around $\vec{n}$ when a photon of frequency ν' from $[\nu',\nu'+d\nu']$ is absorbed in a solid angle $d\Omega'$ around $\vec{n}\,'$. To determine this function Hummer introduced the one atom redistribution function defined in the rest frame of the atom. The probability of absorbing a photon with frequency γ' from $[\gamma',\gamma'+d\gamma']$ and solid angle $d\Omega'$, given in the rest frame of the atom, is defined as:

$$q(\gamma')d\gamma' \frac{d\Omega'}{4\pi} \quad (20)$$

The probability for the emission of a photon with frequency γ in $[\gamma,\gamma+d\gamma]$ if a photon of frequency γ' is absorbed, can be written as:

$$p(\gamma',\gamma)d\gamma \quad (21)$$

The probability of emitting a photon in the direction $\vec{n}$ if the photon is absorbed in the direction $\vec{n}'$, is given by the phase function:

$$g(\vec{n}',\vec{n})\frac{1}{4\pi} \qquad (22)$$

The one atom redistribution function in the rest frame of the atom is consequently given as the product of these three (21-24) probabilities:

$$R_{lu}(\gamma',\vec{n}',\gamma,\vec{n}) = q(\gamma')p(\gamma',\gamma)g(\vec{n}',\vec{n})\frac{1}{4\pi} \qquad (23)$$

As the atoms in the gas move with respect to the lab reference frame, the one atom redistribution function can be expressed in the lab reference frame by using the Doppler transformation:

$$\gamma' = \nu' - \frac{\nu'}{c}\vec{v}_l\cdot\vec{n}' \qquad \text{and} \qquad \gamma = \nu - \frac{\nu}{c}\vec{v}_u\cdot\vec{n} \qquad (24)$$

By averaging the one atom redistribution function over the velocity distribution of the atoms, Hummer determined a global or a many atom redistribution function. The velocity distribution he used to calculate the redistribution function, is the Maxwellian distribution. Neglecting the recoil ($\vec{v}_l=\vec{v}_u$) this finally results in:

$$R(\nu',\vec{n}',\nu,\vec{n}) = \int F(\vec{v})R_{lu}(\nu',\vec{n}',\nu,\vec{n},\vec{v})d\vec{v} \qquad (26)$$

From this equation an angle averaged form can be derived by:

$$\bar{R}(\nu',\nu) = 4\pi\int R(\nu',\vec{n}',\nu,\vec{n})d\Omega' \qquad (27)$$

To describe the frequency redistribution over a wide range of collision rates Huber (1969), Zanstra (1941) and others suggested the following form for the angle averaged redistribution function $\bar{R}(x',x)$:

$$\bar{R}(x',x) = (1-P_c)\bar{R}_{II}(x',x) + P_c\bar{R}_{III}(x',x) \qquad (28)$$

the redistribution function is given as a linear combination of the $\bar{R}_{II}$- and the $\bar{R}_{III}$ redistribution functions. The $\bar{R}_{II}$-function describes the physical situation, where no collisions occur when the atom is in the upper state. The $\bar{R}_{III}$-function describes the situation where many collisions occur. The parameter P_c is the probability that a collision will occur when the atom is in the excited state and is given as $\gamma_c/(A_{ul}+\gamma_c)$ with γ_c the elastic collision rate.

At low collision rate the frequency redistribution will be characterised by the $\bar{R}_{II}$-function. In this situation the excited state is taken to be naturally broadened while the ground state is considered to be infinitely sharp. The probability for absorption and emission in the rest frame of the atom, expressions (20) and (21), can be written as:

$$q(\gamma') = \frac{\Gamma/\pi}{\Gamma^2 + \gamma'^2} \quad \text{and} \quad p(\gamma',\gamma) = \delta(\gamma'-\gamma) \qquad (29)$$

The absorption profile, due to the natural broadening, is described by the Lorentz line form (Wieskoff and Wigner (1930)). As no collisions occur while the atom is in the upper state, the energy is conserved and the frequency of the emitted and absorbed photon are equal. Therefore the emission probability can be represented by a δ-function. Following the previously described steps, Hummer calculated the global redistribution function. The result of this calculation is a very complex function. In many situations (e.g. at high opacities) an analytically simpler angle averaged redistribution function, obtained with (27), is adequate to describe the experiments. In Figure 7 the $\bar{R}_{II}$-function for different absorption frequencies is shown, the frequencies are expressed in Doppler units.

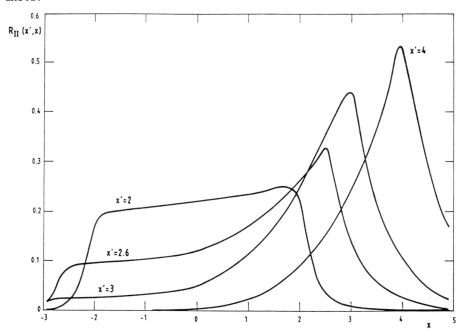

Fig. 7. The angle averaged redistribution function $\bar{R}_{II}(x',x)/L_v(x')$.

It is seen that the probability of emitting a photon in the vicinity of the absorption frequency is very high for large $|x'|$-values. For small $|x'|$-values (say $|x'|<3$) the photon has a large probability of being re-emitted near the line center. This behaviour of the $\bar{R}_{II}$-function led Jefferies and White (1960) to make the following approximation:

$$\bar{R}_{II}(x,x') \cong L_v(x') \left\{ b(x,x')L_D(x) + a(x,x')L_D(x-x') \right\} \quad (30)$$

The first factor represents the absorption, while the second one describes the reemission of the photons. The absorption part is characterised by the Voigt function $L_v(x')$, valid over a large pressure range. The emission part consists of two terms: the first one denotes full frequency redistribution into the Doppler profile $L_D(x)$ when a photon is absorbed in the line center. The second one refers to the re-emission into a shifted Doppler profile $L_D(x'-x)$ when the photon is absorbed in the line wing. This form is complex to use in analytical calculations so further approximations were made. Jefferies and White noted that the $a(x',x)$ and $b(x',x)$-function were complementary and replaced both by a single function $a(x')$, neglecting the x dependence. This $a(x')$-function is equal to 1, when the photon is absorbed in the line wings (say $|x'|>3$) and equals 0 when the photon is absorbed in the line center (say $|x'|<3$). A further approximation was also made by replacing the shifted Doppler line by a δ-Dirac function. This means that when a photon is absorbed in the line wing the photon will be emitted at exactly the same frequency. The $\bar{R}_{II}$-function then becomes:

$$\bar{R}_{II}(x,x') \cong L_v(x') \left\{ (1-a(x'))L_D(x) + a(x')\delta(x-x') \right\} \quad (31)$$

This form is known as the Jefferies-White approximation for the angle averaged $\bar{R}_{II}$ partial frequency redistribution function.

At high collision rate the frequency redistribution is characterised by the $\bar{R}_{III}$-function. In this situation the upper state broadening occurs due to collisions. When the line broadening is treated in the impact approximation the absorption and emission probability in the rest frame of the atom is given as:

$$q(\gamma') = \frac{\Gamma/\pi}{\Gamma^2 + \gamma'^2} \quad \text{and} \quad p(\gamma,\gamma') = \frac{\Gamma/\pi}{\Gamma^2 + \gamma^2} \quad (32)$$

These expressions give rise to the many-atom redistribution function describing this situation. Again this results in a very complex function in which some coherences are present (Jong-Sen Lee 1977). However inclusion of velocity changing collisions yields complete frequency redistribution (Hubeny 1986). As a consequence the angle averaged redistribution function $\bar{R}_{III}$ can be approximated by $L_v(x')L_v(x)$.

With the use of these approximations for the $\bar{R}_{II}$- and $\bar{R}_{III}$-functions Payne derived two criteria to test when the assumption of complete frequency redistribution (CFR) is not valid:

$$P_c < 0.7 \quad \text{and} \quad k(\nu_0)a_v R > 1 \qquad (33)$$

with R the typical distance in the trapping problem and a_v is the Voigt parameter. The first criterion $P_c<0.7$ means that the collision probability must be low as high collision rates disturb the correlation between absorbed and emitted photon frequencies. The second criterion is the requirement that the transport must be in the wings of the Doppler line if correlation between absorbed and emitted photons is to occur. These two criteria must be satisfied at the same time. As an example we applied Payne's criteria with R = 1.8 cm to the $^3P_1 \longrightarrow ^1S_0$ resonance radiation in Ne, Ar, Kr and Xe. The following regions where CFR fails were obtained.

Ne	0.781	< p <	4.0
Ar	0.019	< p <	1.0
Kr	0.0017	< p <	0.8
Xe	0.00056	< p <	0.4

pressures in Torr

A solution to the transport equation (19) was derived by Payne (1974), describing the early time escape, in a special detection geometry following a pencil-like excitation. However for many laboratory experiments a need existed for a solution describing the late time decay. A solution was proposed by Post (1986b), based on the existence of a fundamental mode. The decay rate at late time is given by the following equation:

$$\beta = A_{ul} \int_{-\infty}^{+\infty} dx \, \frac{(1-P_c)L_D(x)+P_c L_v(x)}{1 - (1-P_c)a(x)g(\tau_x,x)} \, \eta(\tau_x,x)$$

(34)

with $\tau_x = k(x)R$ and $g(\tau_x,x) = 1 - \eta(\tau_x,x)$

where $\eta(\tau_x,x)$ is the escape function. The approximation for the escape function at the center of the cylinder $\eta(\tau_x)$ proposed by Huennekens (1989) greatly facilitates the calculation of β. Post compared the theory with his measurements on the radiative transfer of the $Hg(^1P_1)$ 184.9 nm resonance radiation (Post 1986b).

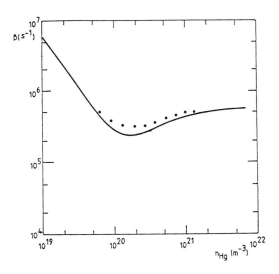

Fig. 8. The experimental and theoretical results of Post(1989a,b) (●, experiment; ——, theory).

Figure 8 shows a good agreement between theory and experiment. However, at the lowest decay rates a discrepancy of about 30% persists. This discrepancy, as Post noted, is probably caused by the Jefferies-White approximation, used in Payne's equation. In this approximation the shifted Doppler profile is replaced by a δ-function preventing the diffusion of the photon in frequency space when the photon is absorbed in the line wings. The photon frequency remains unchanged until the photon reaches the boundary of the gas. As a consequence the theory will give an overestimation of the trapping, in the region where partial redistribution (PRD) is important.

126   *Optogalvanic Spectroscopy*

5. MONTE CARLO SIMULATION FOR RESONANCE RADIATION TRAPPING.

An alternative method of treating resonance radiation imprisonment is by the use of a Monte-Carlo simulation, which has shown to be a more flexible method than the analytical calculations. In MC-simulation we are not restricted to ideal geometries, even a shape as complex as a torus (Denisov 1983) can be treated. Complex redistribution functions can also be used. Generally in Monte-carlo simulation the movement of thousands of photons is traced from their initial emission to the final escape out of the gas. If we consider a situation at low radiation intensities we can treat every photon separately and afterwards, by summing over all photons, the behavior of the whole system can be described. For a detailed review on the application of the MC-simulation to the problem of resonance radiation transport we refer to the paper by House and Avery (1969).

In our laboratory a MC program was developed in a stepwise fashion to simulate the trapping of resonance photons. As a first step and test, we simulated the trapping in the case of complete redistribution for a Doppler broadened profile. Our results should then reproduce the Holstein analytical results. The calculations were carried out for the transport of the $Ar(^3P_1)$ 106.67 nm radiation, trapped in a cylindrical geometry of radius 1.9 cm. The decay of the radiation intensity was simulated and the decay time τ_s was determined by fitting an exponential function at the "late time" results.

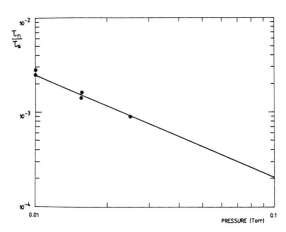

Fig. 9a Comparison between Monte Carlo and analytical calculations of the escape factor $g_1 = \tau_n / \tau_s$ (●, MC-results; ——, Holstein's theory).

Figure 9a shows good agreement between the escape factors determined with our MC-program and Holstein's analytical results. The radial distribution of the excited states at "late times" was also investigated and compared with the analytical calculations by van Trigt (1969); again a good agreement is found (Figure 9b).

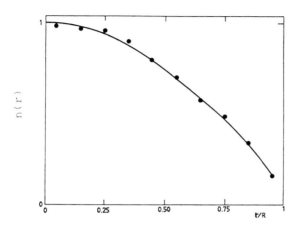

Fig. 9b Comparison between Monte Carlo and analytical calculations of the spatial distribution of the excited states at "late times" (●, MC-results; ——, calculations by van Trigt).

By then we were confident that the simulation was working properly and in a second step the code was extended to include partial frequency redistribution on the basis of the Jefferies-White approximation (Vermeersch 1988). Calculations of the trapping of the Ar($^3P_1$) 106.67 nm radiation confined to a cylindrical geometry of 1.9 cm were performed. The good agreement between the MC-results and the results obtained with Post's formula (34) (see Figure 10) are a strong indication that the eigenvalue method which Post used, is valid. This is because our MC-calculations in which no assumptions are made about the spatial distribution of the excited states, confirm the results obtained with Post's method.

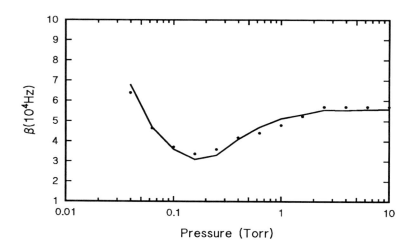

Fig. 10. Comparison between Monte Carlo results and the Post theory. (●, MC-results; ———, Post's theory).

6. EXPERIMENTAL INVESTIGATION OF THE TRANSPORT OF $Ar(^3P_1)$ 106.67 nm AND THE $Xe(^3P_1)$ 146.96 nm RESONANCE RADIATION.

In order to test the validity of the PRD theory, the transport of the $Ar(^3P_1)$ 106.67 nm radiation was studied. We measured the decay of the resonance radiation intensity in the afterglow of an Ar-gas discharge. The discharge tube was a cylinder 35 cm long with 1.8 cm radius. The wavelength selection was performed with a 0.3 m Czerny-Turner monochromator. The selected VUV-light was detected with a SEM 4219 SPIRALTRON™ in single photon counting mode. The single photon pulses were fed to a standard single photon counting circuit. In Figure 11 the measured decay rates are compared with the decay rates according to Post's theory. The natural decay rate τ_n, used in the calculations, is equal to 8.6 ns. This value is taken from the experimental work of Lawrence (1968).

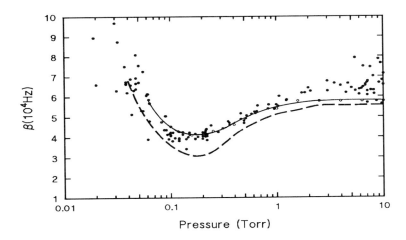

Fig. 11. Measured decay rates at "late times" of the $Ar(^3P_1)$ 106.67 nm radiation as a function of pressure (T=300 K) (●, experiment, ---- Post's theory, ──── present MC-simulation based on the Jong-Sen Lee generators).

It is seen that the theoretical calculations describe the experimental data qualitatively well. At pressures around 10 Torr a discrepancy is found between experiments and theory. This discrepancy can be accounted for by inclusion of collision induced fluorescence into the theory. At pressures around 0.2 Torr where the collision rate is low and where effects of PRD are important, we note a discrepancy of about 30% between experiment and theory. Post (1986b) found a similar discrepancy in his experiments and ascribed this to the Jefferies-White approximation used in his calculations. This led us to re-calculate the decay rates with the use of a MC-program which is not based on the Jefferies-White-approximation. A MC-code based on physically more realistic redistribution functions, generated with algorithms developed by Jong-Sen Lee (1974,1977,1982), was developed. The results of these calculations are included in Figure 11. Better agreement is found between the MC-calculations and the experimental results in the pressure range where the effects of PRD are important.

Similar experiments on the transport of the Xe($^3P_1$) 146.96 nm radiation were undertaken. The excitation of the $^3P_1$ level was achieved by three-photon laser excitation. The typical distance in the experimental cell was 1.8 cm. For further details about the experimental setup we refer to the contributed paper by F.Vermeersch, N.Schoon, E.Desoppere and W.Wieme of this conference. Again the decay of the resonance radiation intensity after the excitation pulse was measured with the use of a single photon counting technique. In Figure 12 the measured decay rates are plotted. Two minima in the decay rates are observed. A first minimum occurs at pressures around 0.015 Torr and a second minimum occurs around 3 Torr. The first minimum is caused by effects of partial frequency redistribution. The second minimum can not be attributed to effects of PRD due to the high collision rate at these pressures is high. At the moment we do not have a consistent explanation of this minimum but we believe that it is probably caused by the quasi-static wing of the emission absorption profile. Further experimental and theoretical investigation on this second minimum is in progress. In Figure 11 we also plotted the results of calculations performed with Post's theory. The natural decay rate used in the calculation was 4.6 ns (Wieme 1979). A qualitative good agreement is found. However a large discrepancy persists in the region where the effects of partial frequency redistribution are important (around 0.015 Torr) i.e. in the region where the Post theory gives an underestimation of the decay rates due to the Jefferies-White-approximation.

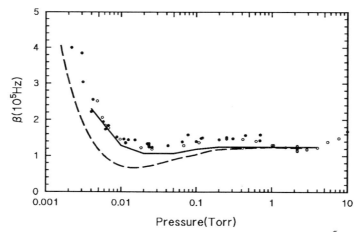

Fig. 12. Measured decay rates at "late times" of the Xe($^3P_1$) 146.96 nm radiation as a function of pressure (T=300 K). ●,○, experiment, ---- Post's theory, ─── present MC-simulation based on the Jong-Sen Lee generators.

We again performed MC-calculations based on the algorithms of Jong-Sen Lee. The results of these calculations are also plotted with the experimental data in Figure 12. A better agreement is found between experiment and theory in the region where PRD is important. The small discrepancy that still persists in this region is probably due to the uncertainty on the natural decay rate τ_n.

8. CONCLUSIONS.

For the calculation of imprisonment of resonance radiation different theories can be used. The choice of the appropriate theory depends on optical depth, on the lifetime of the excited state and on the decorrelating collision rate. At low opacities $k_0R<4$ Milne's modified diffusion theory is quite sufficient to predict the decay rates. At higher opacities the criteria of Payne play an important role in selecting the appropriate theory. When Payne's criteria for PRD are not fulfilled, the Holstein-Biberman theory is valid although care must be taken in choosing the correct line profile. When Payne's criteria are satisfied, the theory of Post can be used to predict the decay rates at "late times". (At "early times" we refer to the theory developed by Payne). However one must keep in mind that, due to the Jefferies-White approximation the results obtained with Post's theory give an underestimation of the decay rate.

A more flexible but more time consuming method of calculating resonance radiation trapping is the MC-simulation. The MC-simulation allows the use of physically more realistic frequency redistribution functions which in turn give a better prediction of the radiation trapping.

Acknowledgement-This work was supported by the Nationaal Fonds voor Wetenschappelijk Onderzoek through Computer grant S6/5-CD.-E180.

References

Anderson J B, Maya J, Grossman M W, Lagushenko R and Waymouth J F 1985 *Phys.Rev.A.* **31** 2968
Biberman L M 1947 *Zh .Eksp. Teor. Fiz.* **17** 416
Blickensderfer R P, Breckenridge W H and Jack Simons 1976
  J. Phys. Chem. **80** 653
Compton K T 1923 *Phil. Mag.* **45** 752
Denisov V I and Preobrazhinski N G 1983
  Opt. Spectrosc. (USSR) **54** 40
Holstein T 1947 *Phys. Rev.* **72** 1212
House L L and Avery L W 1969 *J.Q.R.S.T.* **9** 1579
Huber D L 1969   *Phys. Rev.* **178** 93
Hubeny I and Cooper J 1986 *Astrophys. J.* **305** 852

Huennekens J and Colber T 1989 *J.Q.R.S.T.* **41** 439
Hummer D G 1962 *M.N.R.A.S.* **125** 21
Irons F E 1979a *J.Q.R.S.T.* **22** 1
Irons F E 1979b *J.Q.R.S.T.* **22** 21
Irons F E 1979c *J.Q.R.S.T.* **22** 37
Jefferies J and White O 1960 *Astrophys. J.* **132** 767
Jong-Sen Lee 1972 *Astrophys. J.* **192** 465
Jong-Sen Lee 1974 *Astrophys. J.* **218** 857
Jong-Sen Lee 1982 *Astrophys. J.* **255** 303
Lawrence G M 1968 *Phys. Rev.* **175** 40-4
Milne E A 1926 *J. London Math. Soc.* **1** 40
Preobrazhinski N G, Senina A V and Paskar S V 1968
   Sov. Phys. J. **11** 23
Payne M G and Cook J D 1970 *Phys. Rev. A* **2** 1238
Payne M G, Talmage J E, Hurst G S and Wagner E B 1974
   Phys. Rev. A **9** 1050
Phelps A V 1958 *Phys. Rev.* **110** 1362
Post H A 1986a *Phys. Rev. A* **33** 2003
Post H A, van de Weijer P and Cremers R M M 1986b
   Phys. Rev. A **33** 2017
Romberg A and Kunze H-J 1988 *J.Q.S.R.T.* **39** 99
Stewart R S, McKnight K W and Hamad K I 1990
   j. Phys. D: Appl. Phys. **23** 832
van Trigt C 1969 *Phys. Rev.* **181** 97
Vermeersch F, Fiermans V, Ongena J, Post H A and Wieme W 1988
   Phys. Rev. B: At. Mol. Opt. Phys. **21** 1933
Weisskoff V and Wigner E 1930 *Zh. f. Phys.* **63** 54
Wieme W and Vanmarcke M 1979 *Phys. Lett.* **72A** 215
Zanstra H 1941 *M.N.R.A.S.* **101** 273

Experimental investigation of the imprisonment of the ($^3P_1-^1S_0$) 146.96 nm resonance radiation in Xenon

F Vermeersch, N Schoon, E Desoppere and W Wieme

Laboratorium voor Natuurkunde, Rijksuniversiteit Gent,
Faculteit Toegepaste Wetenschappen, Rozier 44, 9000 Gent, Belgium

ABSTRACT: Accurate measurements of the decay of the 146.96 nm resonance radiation in Xe are presented for pressures ranging from 0.002 to 10 Torr. A comparison with different theories is made, clearly demonstrating the inability of the complete frequency redistribution theory to describe the experimental results in the whole pressure region. A better agreement is found with the analytical solutions based on a partial frequency redistribution theory. Monte-Carlo calculations, using the partial frequency redistribution function generators of Jong-Sen Lee, have been carried out, reproducing the experimental results in a satisfactory way.

1. EXPERIMENTAL SETUP

The $^3P_1$ level of xenon is populated by means of a non-resonant three-photon excitation process. The excitation source is a pulsed dye laser (Quantel TDL III), which is transversely pumped by a Nd-Yag laser (Quantel YG 481) operating at 10 Hz. A block diagram of the experimental setup can be found in Figure 1. The tunable laser is operated with the dye coumarine 440 in the wavelength region centered around 440.8 nm. The maximum output is ± 10 mJ with a pulse length of about 10 ns (FWHM). The dye laser beam is focused into a cell containing xenon at a well defined pressure. The filling pressure is measured with a baratron (100 Torr head), which has an accuracy of ± 1% reading for pressures greater than 10^{-2} Torr and ± 10% reading for pressures of the order of 10^{-3} Torr. Careful application of high vacuum techniques preclude quenching of the resonance radiation by residual gas impurities. At one end the cell is fitted with a magnesium fluoride window, which is transparent for the 147 nm resonance radiation. This radiation is detected by a solar blind photomultiplier (Thorn EMI G26E314LF), which was selected for low dark counts.

© 1991 IOP Publishing Ltd

One of the great advantages of the experimental setup resides in the ability to populate selectively the $^3P_1$ level, so that no wavelength selection is needed in the detection system. This contrasts with gas discharge excitation, where several decay channels are possible, which require a wavelength selection. On the other hand, due to the small cross sections of multiphoton processes, the laser beam has to be focused to obtain a measurable

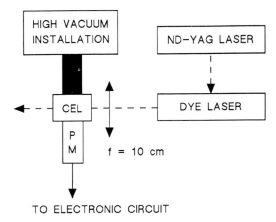

Figure 1 : block diagram of the experimental setup

signal, causing a shift of the $^3P_1$ level by the Stark effect. This shift varies in time during the laser pulse and is dependent upon the spatial position in the focus. This implies that the excitation efficiency will have a spatial as well as a temporal dependency during the laser pulse (± 10 ns) for a given exciting wavelength. However, this will not influence the *late time* decay of the resonance radiation (our measurements are limited from 3 to 40 μs after the laser pulse): at these *late times* the decay is described by its fundamental mode (Holstein 1947, Post 1986), which is independent upon the starting distribution of the excited atoms.

A first set of results has been obtained by measuring the decay of the resonance radiation using a single photon counting technique, as described by Davis (1970). The three major components of this system are a time to pulse height converter, a multichannel analyzer and a logic gate. When more than one photon is detected during the measuring time the gate will be closed, rejecting the cycle. Only in the case that one photon is detected during the measuring time, the count will be added to the multichannel signal. This logic gate is needed, because this is the only possibility to measure the decay with this system in a correct way: the necessity of focusing the laser beam increases the probability of detecting more than one photon during the measuring time. The logic gate is used to retain the measurements during which only one photon is detected. The probability of detecting one single photon is optimized through attenuation of the laser beam. This detection technique entails long registration times.

Another set of measurements has been carried out using a fast multichannel scaler (Stanford Research Systems), which is directly coupled to the photomultiplier. With this instrument the registration time is reduced with a factor 10 and much better statistics are obtained.

2. RESULTS

In Figure 2 two typical examples of the measured decay of the resonance radiation at 0.0056 and 0.0257 Torr are shown. In all cases the decay could be fitted with a function of the form $A\,e^{-\beta t} + B$, which proves that at late times the fundamental mode has been reached for the decay. The coefficient B takes into account the background and is less than one count per second, due to the small dark current of the photomultiplier.

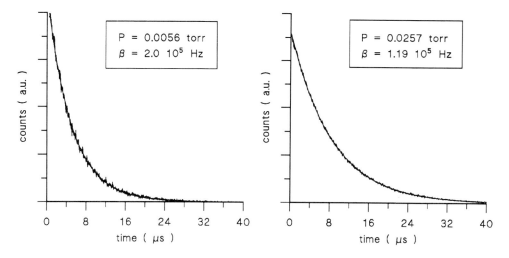

Figure 2 : two typical examples of the measured decay of the resonanace radiation

In Figure 3 the fitted decay rates are plotted as a function of the pressure. The results marked with circles are obtained with the single photon counting technique, while the ones marked with triangles are obtained with the multichannel scaler. The low scatter of the latter is clearly visible. The statistical error in a single value of β is very small, but if the reproducibility of the measurements is taken into account, the error in β is estimated to be 10% for the measurements with the single photon counting technique and ± 5% for the ones with the multichannel scaler.

3. DISCUSSION

The accuracy of our measurements allows us to to compare these results with the different theories describing the trapping of resonance radiation.

In a first step, a comparison is made with the analytical solutions proposed by Holstein (1947, 1951), assuming complete frequency redistribution, i.e. no correlation exists between the frequency of the absorbed and emitted photon. Holstein obtained the following equation to describe the evolution of the population of the excited atoms:

$$\frac{\partial}{\partial t} N_u(\vec{r},t) = - A_{ul} N_u(\vec{r},t) + A_{ul} \int_V N_u(\vec{r}',t) G(\vec{r}',\vec{r}) dV'$$

$N_u(\vec{r},t) d\vec{r}$ is the number of excited atoms at time t in $d\vec{r}$ at $\vec{r}$. The first term on the right hand side of the equation describes the spontaneous emission, with A_{ul} the Einstein coefficient, while the second term describes the increase of $N_u(\vec{r},t)$ due to absorption of photons, which are emitted at a place $\vec{r}'$. $G(\vec{r}',\vec{r}) d\vec{r}$ is the probability that a photon emitted at $\vec{r}'$ will be absorbed in $d\vec{r}$ at $\vec{r}$. Holstein showed that at late times the evolution of $N(\vec{r},t)$ is described by a single exponential decay at every spatial point:

$$N_u(\vec{r},t) = c\, n(\vec{r})\, e^{-\beta t}$$

The decay rates for a cylindric geometry in the case of Doppler broadening of the line and in the case of natural and pressure broadening are given by the following formulas:

Doppler broadening : $\beta = 1.6\, A_{ul} \left(k_0 R \sqrt{\pi \ln\left[k_0 R \right]} \right)^{-1}$

natural + pressure broadening : $\beta = 1.125\, A_{ul} \left(\sqrt{\pi k_0 R} \right)^{-1}$

R is the radius of the cylinder and k_0 is the absorption coefficient for the central line frequency ν_0. If these analytical solutions are compared with our results (see Figure 3), it is clear that the assumption of complete frequency redistribution is only valid at low pressures, where the transport takes place in the central core of the Doppler profile, and for high pressures, where frequency decorrelating collisions play an important role in the imprisonment of the resonance radiation. In the region, where the transport takes place in the wings of a Doppler profile, which is moderately broadened by collisions, the complete frequency redistribution theory is unable to explain the experimental results.

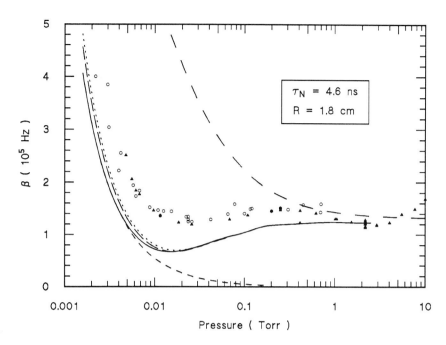

- ○ measurements with single photon counting technique
- ▲ measurements with multichannel scaler
- – – – complete frequency redistribution (Holstein) : Doppler broadening
- — — complete frequency redistribution (Holstein) : natural and pressure broadening
- ——— partial frequency redistribution (Post) : analytical solution
- — — influence of the isotopic effect
- ······· influence of the isotopic effect and of diffusion

Figure 3 : comparison between measurements and analytical solutions

In a second step the results can be compared with the analytical solution, proposed by Post (1986), who used the transport equation of Payne (1974), based on partial frequency redistribution:

$$\frac{\partial}{\partial t} N_u(\vec{\rho},x,t) = -A_{ul} N_u(\vec{\rho},x,t) + S(\vec{\rho},x,t) - A N_u(\vec{\rho},x,t)$$

$$+ A_{ul} k_0 R \pi^{1/2} \int_{-\infty}^{+\infty} dx' \int_V d\vec{\rho}' \; N_u(\vec{\rho}',x',t) \; \bar{R}(x',x) \; \frac{e^{-k(x') R |\vec{\rho}'-\vec{\rho}|}}{4\pi \; |\vec{\rho}'-\vec{\rho}|^2}$$

$$\vec{\rho} = \frac{\vec{r}}{R} \quad ; \quad x = \frac{\Delta\nu}{\Delta\nu_D} = \frac{\nu - \nu_0}{\nu_0} \frac{c}{\bar{v}} \quad ; \quad \bar{v} = \sqrt{2kT/m} \quad ; \quad k(x) = k_0 \sqrt{\pi} \; L_v(x)$$

$N_u(\vec{\rho},x,t)\,d\vec{\rho}\,dx$ is the number of excited atoms in $d\vec{\rho}$ at reduced distance $\vec{\rho}$ at time t, which emit a photon into the frequency interval dx at x. Again the first term on the right hand side of the equation describes the spontaneous emission; the second term describes the increase of $N_u(\vec{\rho},x,t)$ due to processes other than absorption; the third term describes the deexcitation by processes other than emission and the fourth term describes the increase due to absorption of photons with frequency x' emitted at a place $\vec{\rho}'$. The partial frequency redistribution is included in the factor $\bar{R}(x',x)$, defined such that $\bar{R}(x',x)\,ds\,k_0\,\pi^{1/2}\,dx$ describes the probability of a photon with frequency x' being absorbed while traversing a distance ds and being reemitted into dx at x. An approximation of this angle averaged partial frequency redistribution function, suggested by Jefferies and White (1960), has the following form:

$$\bar{R}(x',x) = L_v(x')\left[(1-P_c)\left((1-a(x'))L_D(x) + a(x')\delta(x'-x)\right) + P_c L_v(x)\right]$$

$L_D(x)$ describes the Doppler line profile; $L_v(x)$ represents the Voight profile; P_c is the probability that a frequency decorrelating collision occurs; a(x) is a function, which reaches the value 1 for frequency differences much greater than three Doppler widths and reaches the value 0 for frequencies in the central core of the line profile. This means that a fraction P_c, due to frequency decorrelating collisions, will be redistributed according to a Voight profile $L_v(x)$, i.e. complete frequency redistribution. The other fraction $(1-P_c)$ will be redistributed according to the frequency of the absorbed photon: if the photon is absorbed in the central core, than the photon will be emitted according to a Doppler profile $L_D(x)$, which is again complete frequency redistribution. However, if the photon is absorbed in the wing of the line profile, the photon will be emitted at exactly the same frequency. In this case, a correlation between the frequencies of the absorbed and emitted photon exists. Post argues that the decay of the resonance radiation at late times is governed by the fundamental mode. For the decay rate he found the following expression, where η is the time independent escape function:

$$\beta = A_{ul}\int_{-\infty}^{+\infty}\frac{(1-P_c)L_D(x) + P_c L_v(x)}{1 - (1-P_c)a(x)(1-\eta(\vec{\rho},\tau_x,x))}\eta(\vec{\rho},\tau_x,x)\,dx \qquad \tau(x) = k(x)R$$

From Figure 3 it is clear that above mentioned solution describes our results better.

However, in the central pressure region a discrepancy of 30 % to 50 % still remains. This can be ascribed to the use of the delta function approximation, introduced by Jefferies-White, to describe the correlation between the frequencies of the absorbed and emitted photon, when the first is absorbed in the wing of the profile. By using the delta function the frequency of the emitted photon can not shift towards outer lying optically thinner frequencies, so that the photon is trapped longer in the gas, resulting in a slower decay.

Post refined the above mentioned formula to describe the influence of the hyperfine structure on the decay rate:

$$\beta = A_{ul} \int_{-\infty}^{+\infty} \frac{\sum_i (1-P_{c_i}) d_i L_D(x) + \sum_i P_{c_i} d_i L_V(x)}{1 - \sum_i (1-P_{c_i}) a_i(x) (1-\eta(\vec{\rho},\tau_x,x)) \frac{\tau_{x_i}}{\tau_x}} \eta(\vec{\rho},\tau_x,x) \, dx, \quad \tau_x = \sum_i \tau_{x_i}$$

d_i is the natural abundance of the different isotopes. Figure 3 shows that in the case of xenon this effect is only noticeable at low pressures. At higher pressures this effect is completely dominated by the line broadening, due to frequency decorrelating collisions.

A further refinement consists in adding the term $-D\nabla^2 N_u(\vec{r},t)$ to Holstein's transport equation to describe the influence of diffusion (Phelps 1958). This results in a small supplementary term for the decay rate:

$$\beta^D = \beta^H + \frac{39.6 \text{ Torr Hz}}{p \text{ (Torr)}}$$

Again, it is clear from Figure 3 that the influence of diffusion is neglectable in the case of xenon and is certainly smaller than the experimental errors in β in the low pressure range.

Finally, Monte-Carlo calculations have been carried out. The calculations, using the Jefferies-White approximation, showed the same discrepancy as the analytical calculations (Vermeersch 1987). However, as shown in Figure 4, Monte-Carlo calculations, performed at our laboratory, using the physically more realistic partial frequency redistribution function generators of Jong-Sen Lee (1974, 1977, 1982), in which no approximations are made for the redistribution function $\bar{R}(x',x)$, describe our results much better.

140   *Optogalvanic Spectroscopy*

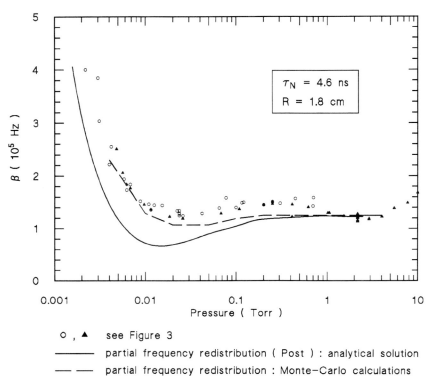

○ , ▲   see Figure 3
─────── partial frequency redistribution (Post) : analytical solution
── ── partial frequency redistribution : Monte-Carlo calculations

<u>Figure 4</u> : comparison between measurements and Monte-Carlo calculations

Due to the accuracy of the measurements with the multichannel scaler, an unexpected second minimum is found in the pressure region around 3 Torr, a feature, which is up till now not observed in other experiments. Work is in progress to include in the present theory the quasi-static wing of the line profile, which we think could account for this phenomenon.

This research was supported by the F.K.F.O. grant 2.0020.90 and the N.F.W.O. computer grant S6/5-CD.-E180.

REFERENCES

Davis C C and King T A 1970 J.Phys.A:Gen.Phys. **3** 101
Holstein T 1951 Phys.Rev. **83** 1159
Holstein T 1947 Phys.Rev. **72** 1212
Jefferies J and White O 1960 Astrophys.J. **132** 767
Jong-Sen Lee 1974 Ap.J. **192** 465
Jong-Sen Lee 1977 Ap.J. **218** 857
Jong-Sen Lee 1982 Ap.J. **255** 303
Payne M G, Talmage J E, Hurst G S and Wagner E B 1974 Phys.Rev. **A9** 1050
Phelps A V 1958 Phys.Rev. **110** 1362
Post H A 1986 Phys.Rev. **A33** 2003
Vermeersch F, Fiermans V, Ongena J, Post H A and Wieme W 1988 J.Phys.B: At.Mol.Opt.Phys. **21** 1933

Inst. Phys. Conf. Ser. No. 113: Section 5
Paper presented at International Meeting on Optogalvanic Spectroscopy, Glasgow, 1990

Aspects of trace analysis using resonance ionization mass spectroscopy and some applications

K W D Ledingham

Department of Physics and Astronomy, University of Glasgow, Glasgow G12 8QQ, Scotland, UK

ABSTRACT: Resonance Ionisation Mass Spectroscopy (RIMS) is a unique, ultra sensitive analytical technique which can detect down to the level of a few atoms or molecules. In this paper the concepts, methodology and instrumentation associated with RIMS will be discussed in detail. In addition certain applications of RIMS and its associated technology will be investigated, with particular reference to depth profiling of semiconductor materials, detection of molecules in supersonic jets, carbon cluster formation in the laser ablation process and finally resonant laser ablation.

1. PREAMBLE

It is probably worthy of comment why Resonance Ionisation Mass Spectroscopy (RIMS) should be included in a conference dealing with Optogalvanic Spectroscopy. However, when the details of the two processes are carefully compared, marked similarities become obvious, a point which has already been noted (Hurst and Payne 1988a). In a modern optogalvanic spectroscopy arrangement a laser is passed through a gas discharge in a hollow cathode lamp. When the laser is tuned to a transition between two levels of atoms or ions in the discharge, the population densities of these levels change. Because the ionisation probability from these two levels is different, the change in population density causes a change in the discharge current in the hollow cathode lamp which is detected. In optogalvanic spectroscopy, after photon absorption, the ionisation step in the discharge is caused by either electric impact, atomic collisions or photoionisation. Very similar processes occur in Resonant Ionisation Mass Spectroscopy although not normally in a gas discharge.

2. INTRODUCTION

Resonance Ionisation Spectroscopy (RIS) is a technology which was developed during the seventies, principally by Letokhov and his co-workers in Moscow (Letokhov 1986) and Hurst and Payne (1988b) at Oak Ridge in the USA. Early on, it was recognised that RIS was a very sensitive technique which could detect down to the single atom level (Hurst et al. 1977), however it took several more years - until the early eighties - before the methodology was sufficiently developed that the ultra sensitive analytic technique of Resonance Ionisation Mass

© 1991 IOP Publishing Ltd

Spectroscopy evolved (RIMS) (Bekov and Letokhov 1983 , Hurst and Payne 1988a , Fassett and Travis 1988 and Lubman 1987) evolved.

Although the samples to be analysed can be in solid, liquid or gaseous states, the preferred form is solid and both stable and radioactive material can be assayed. If the sample is a solid then analysis using a modern RIMS arrangement consists of three steps (fig. 1).

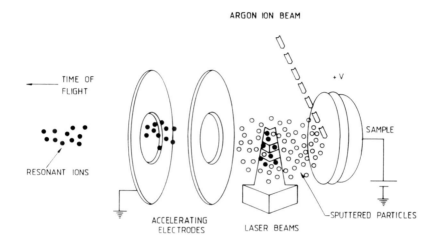

Fig. 1   The three steps involved in RIMS:

    a)   sputtering/ablation
    b)   resonant laser ionisation
    c)   ion extraction and mass indentification

a) The sample is gasified by either ion sputtering or laser desorption to form a plume of principally neutral atoms or molecules.

b) The neutral atoms or molecules in the plume are resonantly ionised using one or more laser beams normally pulsed and broadband. For atoms, this step is elementally very selective with isotopic selectivity being carried out in the next step (c). Isobaric interferences which plague conventional mass spectroscopy are largely eliminated at this stage.

c) Finally the ions created are mass analysed, normally using a time-of-flight mass spectrometer which complements ideally the pulsed aspect of the resonant ionisation step.

Each of these steps will be described in the next section (Procedure).

3. PROCEDURE

3.1 Ion Sputtering and Laser Desorption

Either of these processes for creating neutral atoms or molecules has pros and cons. Ion sputtering is preferred if profiling is involved, i.e. determining the concentration of a particular element as a function of depth below the surface, and is invariably used for analysing semiconductor materials. On the other hand laser desorption is a soft ablation procedure for liberating molecules from a surface intact, i.e. with minimal fragmentation which often occurs with ion sputtering. Hence laser desorption is popular in the analysis of organic molecules especially when preserving the parent mass peak is important, although it should be pointed out that laser desorption is much less well characterised.

3.2 Ion Sputtering

The most important aspect of ion sputtering is to maximise the sputtering yield, i.e. the number of neutral atoms created from the sample per incident ion. The yield is a complicated function of many parameters among which are the incident ion's mass and energy as well as the incident angle, i.e. the angle between the beam and the normal to the sample. A comprehensive treatment of ion sputtering is given in Benninghoven et al. (1987) dealing with Secondary Ion Mass Spectrometry (SIMS), a technique which has been established for many years. SIMS employs a similar technology to RIMS but analyses the sputtered ions created by the incident beam rather than the neutrals - a point to which we shall return. Figs. 2 and 3 show how the sputtering yield varies as a function of the incident ion's mass, energy (Townsend et al. 1976) and angle of incidence (Oeschner1973). Clearly an argon ion gun with 5-10 keV energy fitted at an angle of incidence of about 60-70° to the normal is a sensible instrument choice for sputtering and maximises the sputtering yield at about 5. Typically pulsed ion guns have currents of about a few μ amps in pulses of about 1μ sec. duration, and hence the number of sputtered neutrals per pulse per μ amp is $\sim 3 \times 10^7$.

3.3 Laser Desorption (Ablation)

When a laser beam strikes the surface of a sample, vaporisation is referred to as desorption if the photon flux is less than about $10^8 Wcm^{-2}$ or ablation if the flux exceeds this value. Mass ablation coefficients in $kg\ cm^{-2}s^{-1}$ have been determined experimentally by Fabbro et al. (1982) and Goldsack et al. (1982) from which one can calculate the number of ablated atoms from the sample surface (assuming the atoms are created free). These however only refer to ablation at fluxes above $10^{12} Wcm^{-2}$ and it is not certain whether the analytical expressions can be used down to lower fluxes. Above flux levels of $10^8 Wcm^{-2}$ microscopic craters and measurable material removal exist. Very little work on measuring neutral emission rates has been carried out at flux levels below damage threshold but Arlinghaus et al. (1989) have determined the neutral yield of Zn atoms from ZnS crystals desorbed at 308nm to vary between

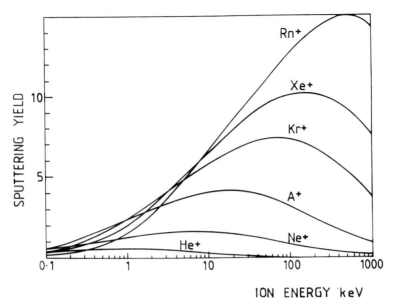

Fig. 2 Calculated sputtering yields for aluminium as a function of ion energy and ion mass *(Townsend et al. 1976)*
Reproduced by permission of the authors and Aacademic Press

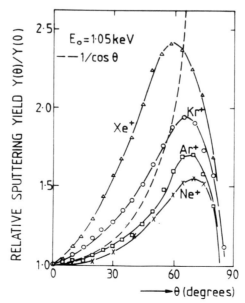

Fig. 3 Relative sputtering yields ($\theta°/0°$). For Ne^+, Ar^+, Kr^+ and Xe^+ as a function of angle of incidence (θ) to the normal *(Oechsner 1973)*
Reproduced by permission of the authors and Springer-Verlag

10^8 - 10^{12} atoms per pulse as the laser fluence was varied between 20 and 80 mJ/cm^2.

It is clear however that a typical laser pulse desorbs many more atoms than a typical ion gun pulse. Furthermore Kelly and Dreyfus (1988) have shown that the angular dependence of a laser induced plume is much more forward peaked than the cos θ dependence expected from ion sputtering. This is likely to ease the difficulty of overlapping the neutral plume and the ionising lasers.

4. RESONANT IONISATION OF NEUTRALS

Resonant ionisation spectroscopy is a photoionisation method by which the neutral atoms in the plume above the sample are ionised by the absorption of two or more photons from tunable dye lasers. One or more of the absorbed photons energetically match transitions between quantum states of an atom or molecule and hence the process is resonant. Hurst and Payne (1988a/b) have proposed five basic ionisation schemes according to the relative energy position of the intermediate states to the continuum. These are shown in fig. 4. In the first scheme a level exists at an energy more than $1/2$ IP (Ionisation Potential) and hence the atom can be ionised by a resonant two photon process (same colour). In scheme two the output of the dye laser (λ) must be frequency doubled ($\lambda/2$) to excite the atom and then ionised by a photon (λ) from the original beam. The other three schemes involve the absorption of three photons. Fig. 5 shows a diagram of the periodic table with one of the five schemes being ascribed to each element. It must be emphasised that these are only suggested schemes for ionising a particular element and a study of the atomic energy level tables, e.g. C E Moore (1948), will suggest many more. It can be seen from fig. 5 that many elements can be ionised using a single dye laser which has a frequency doubling capacity including most of those important to the semiconductor industry. For complete elemental coverage, one requires a large pump laser (Excimer or Nd:YAG) and two dye lasers, one of which is frequency doubled. The laser linewidth typically used in RIMS measurements is between 0.1 and 1.0cm^{-1} and hence all isotopes are ionised simultaneously.

To maximise the number of ions formed, the ionisation process should be saturated, i.e. each neutral atom in the laser beam will be ionised. Fig. 6 shows typical atomic excitation and ionisation cross sections. In 6a the excitation step can be saturated with laser fluences of about 1μJ/cm^2 (Singhal et al. 1989) while the ionisation step requires fluences of about 100mJ/cm^2. To reach such fluences the output of most commercially available dye lasers must be moderately focused which has the disadvantage of reducing the interaction volume and hence the number of atoms which can be analysed. Two other ionisation procedures have been suggested by Bekov and Letokhov (1983) which alleviate this problem and are shown in fig. 6(b) and (c). In (b) the atom is excited to a Rydberg level close to the continuum and can be ionised with 100% efficiency by the application of a pulsed field of about 10^4V/cm. The other method is to ionise the atom via autoionisation states (c). The rate limiting step in (b) and (c) has a cross section some two orders of magntiude larger than in (a) and hence requires much

146 Optogalvanic Spectroscopy

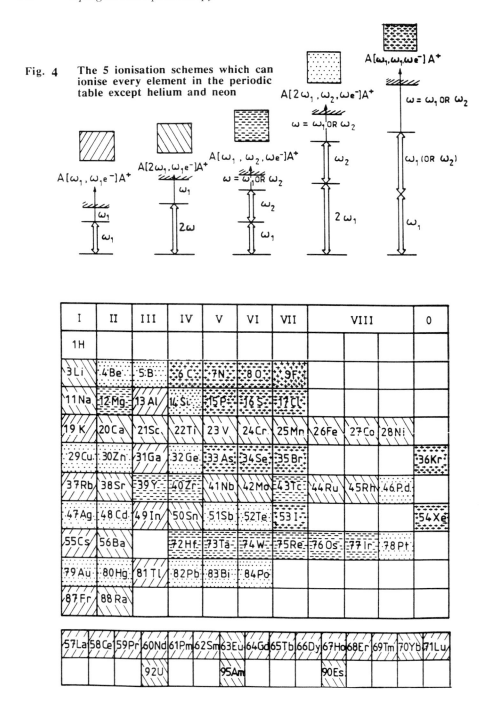

Fig. 4 The 5 ionisation schemes which can ionise every element in the periodic table except helium and neon

Fig. 5 Periodic table with an appropriate scheme for each element

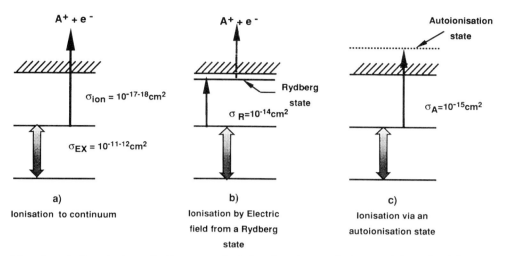

Fig. 6 **a)** An electron in its ground state absorbs a photon and is raised to an excited state. Ionisation by absorption of a second photon has a small cross section.

b) The atom is excited to a Rydberg state and is finally ionised by a pulsed electric field with high efficiency.

c) The final step is to an autoionisation state with a large cross section.

lower fluences to reach saturation.

Finally to maximise the overlap between the sputtered/ablated atoms and the ionising lasers, the RIS beams should be unfocused and pass as close to the sample as possible. The timing between the sputtering/ablation and ionising pulses is also important with the latter being triggered between 0-2µs after the completion of the former (Kimock et al. 1984).

5. ION DETECTION

After ionisation, the ions are extracted and passed into a time-of-flight (TOF) mass spectrometer. Although quadrupole and magnetic sector mass spectrometers have been used in RIMS, TOFs are usually employed since all masses can be analysed in one pulse and also the transmission through the spectrometer is generally superior to other types. The ion extraction optics and a time-of-flight mass spectrometer of reflectron type is shown in fig. 7 (Towrie et al. 1990). The principal factor which limits the resolution of a conventional TOF spectrometer is the spread of initial ion energies in the sputtering/ablation process. This spread of ion energies can be compensated using a reflectron TOF (Mamyrin et al. 1973) in which higher energy ions penetrate deeper into an electrostatic ion reflector and hence experience a longer flight time than ions of a lower energy. Thus the spread in initial ion energies which result in a spread of the arrival times of the ions at the detector is considerably reduced. The FWHM resolution of the

148   *Optogalvanic Spectroscopy*

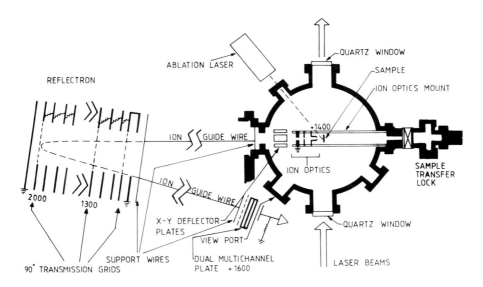

Fig. 7   The Glasgow RIMS instrument showing reflectron and an ion guide wire to increase transmission.

above system is about 1000 for ions of about 40amu. Finally the ions are detected by a Galileo double microchannel plate detector with a 0.2ns rise time. The flight time for ions of about 40amu is typically about 50µs for the overall drift length of 3m operated with extract voltages at about 2kV. A thin wire (0.005cm in diameter) follows the ion flight path through the flight tube providing an electrostatic guide for the ions increasing the transmission of the mass spectrometer. This wire is normally operated at about -10V. The sample chamber is pumped by an oil diffusion pump fitted with a cold trap and a titanium sublimation pump having a total pumping speed of over 800L/sec which can maintain a base pressure of less than 10^{-9} torr.

The data acquisition system measures and stores mass spectra and laser probe energies on a pulse to pulse basis and is described in greater detail elsewhere (Towrie et al. 1990). A LeCroy 2261 transient recorder coupled to a IBM PC AT forms the centre of the system. Ion signals from the detector are digitised by the transient recorder which provides 640 time channels (11 bit resolution) each of 10µs width. This provides time spectra of width 6.4µs which can be delayed at will by a Stanford precision pulse generator to give complete mass coverage.

6.   TYPICAL ANALYTICAL RESULTS

The sensitivity of the instrument can be tested using NBS standard materials. NBS standard coals and ashes (SRM 1632a, 1633, 1645) contain many elements at ppm concentrations (Gladney 1980) as well as having a difficult matrix problem with the ablation of a range of carbon clusters and hence is a stringent test of the capability of the technique. The sample stub was 2.5cm in diameter with an inset cm in diameter and 2mm deep in the middle of the stub. A well homogenised pellet of 50% SRM 1632a and 50% high purity pelletable graphite to make the sample conducting was placed in the inset. Conducting samples reduce space charge problems which limit resolution. Fig. 8 shows a Rb (IP 4.18eV) signal at 15ppm concentration. An unfocused laser beam of wavelength 532nm and fluence 1mJ/mm^2 was used for ablation purposes. The neutral Rb atoms were ionised using a two step process: a photon from stilbene 420 dye to excite (420.7nm, 120µJ focused) and a doubled photon Nd:YAG laser (532nm, 7mJ unfocused) to augment the ionisation step. The two laser beams passed through the interaction volume colinearly and were triggered to ionise the neutral plume about 1µs after the ablation pulse. The broad peaks to left and right of the Rb peaks are C_7 and C_8 clusters produced by unsuppressed ablation ions. The background level in fig. 8 corresponds to a minimum detection level of Rb in SRM 1632A coal of < 1 ppm.

Elimination of the unsuppressed ablation ions can be carried out using a procedure reported by Pellin et al. (1986). In this approach emission of secondary ions from the sample ceases very soon after the completion of the ablation pulse. By pulsing the optics during the ablation cycle, positive secondary ions are forced to remain at the sample surface.

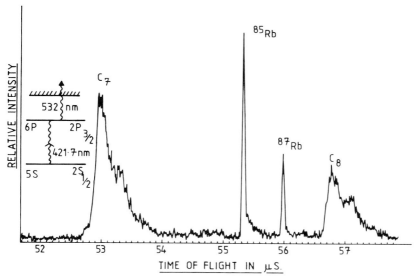

Fig. 8   Rb signal at 15ppm concentration in an NBS coal sample SRM1632a. The broad peaks C_7 and C_8 are carbon clusters produced in the ablation process. The background corresponds to a concentration of less than 1 ppm.

Calcium (IP 6.11eV) was chosen as another very suitable element for test purposes since it has a range of isotopes with large abundance ratios: ^{40}Ca (96.94), ^{42}Ca (0.647), ^{43}Ca (0.135), ^{46}Ca(0.004) and ^{48}Ca (0.187). High purity calcium carbonate was homogeneously mixed with high purity graphite in the mass ratio of 1:4 which resulted in ^{46}Ca being present at abundance levels of about 3ppm. The sample was ablated with a laser power of 1mJmm^{-2} and then reasonantly ionised via a three-photon ionisation process with an output level of 2mJ highly focused. The two-photon excitation transition $^1S_0$-$^1D_2$ at 2 x 536nm was used followed by the absorption of a third 536nm photon to ionise. Fig 9 is a typical calcium mass spectrum accumulated from the addition of several thousand laser shots and shows the ^{46}Ca peak with an abundance level of about 3ppm.

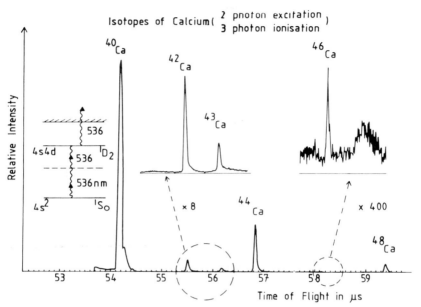

Fig. 9 Resonant ionisation spectra of the Ca isotopes 40, 42, 43, 44, 46 and 48 using a three-photon process. The spectrum is the summation of 10^4 shots and the ^{46}Ca$^+$ ion peak has an isotopic abundance of 3 ppm.

In this paper only analyses at ppm trace concentrations have been discussed due to the fact that reliable standards at sub ppm levels are difficult to obtain. In a recent paper however Arlinghaus et al. (1990) have analysed indium in silicon down to 8ppb and sensititivies have been reported even at parts in 10^{12} for specific elements (Bekov et al. 1985).

7. APPLICATIONS OF RIMS

7.1 Depth Profiling of Layered Structures in Semiconductor Materials

In fig. 1 it was shown that if an ion beam or laser beam sputters or ablates the sample both ions and neutrals are emitted. The detection of the ions is the basis of the well characterised and mature technique, Secondary Ionisation Mass Spectroscopy (SIMS). The SIMS technique has two disadvantages - firstly, SIMS analyses the ion fraction of the ejected particles which is often 10^{-2} or less as estimated by the Saha equation (Fürstenan 1981) and secondly the analyte fraction is a sensitive function of the sample matrix. This becomes an even more complicated problem if the matrix is a changing layered structure. RIMS does not have these disadvantages and is therefore potentially a more quantitative ultra trace analytical technique. Fig. 10 shows a comparison between a SIMS and RIMS depth profile to detect beryllium in a layered structure of GaAs and AlAs (Downey et al. 1990). Alternative layers of GaAs and AlAs, 100nm thick, were grown on an GaAs substrate at 400°C and doped with Be to nominal uniformity at approximately $2 \times 10^{19} \text{cm}^{-3}$. It is known that under these growth conditions Be is ten times more soluble in GaAs than in AlAs. It can be seen that the RIMS profile represents the layered structure better than SIMS. Moreover, the Be spikes at the interfaces are of equal widths in the RIMS profile whereas in the SIMS data the widths depend on the ordering of the layers. Other matrix independent RIMS data has also been reported by Arlinghaus et al. (1990).

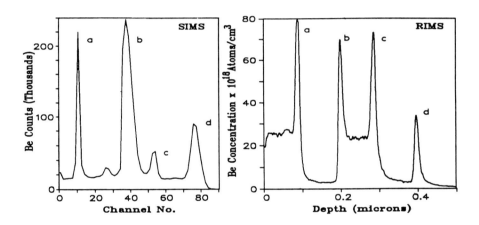

Fig. 10   Depth profiles for Be doped $(GaAs/AlAs)_{x\,2}$ sample.   The SIMS profile was obtained using O^+_2 ion sputtering. The top layer and that between b and c are GaAs and a - d mark the interfaces.
Reproduced by permission of the authors and the Institute of Physics

7.2 Laser Molecular Spectroscopy in Supersonic Beams (jets)

Resonant Ionisation Mass Spectroscopy is a description more associated with the detection of atoms which usually have a simple set of energy levels. The same techniques can be used with molecules but molecular structure is considerably more complex since each electronic level has an associated set of vibrational and rotational levels. Unlike atoms, the bound-bound transitions in molecules have cross sections similar in magnitude to the bound-continuum transitions. A multiphoton process in molecules where the laser is tuned to an intermediate electronic state is normally called resonance enhanced multiphoton ionisation (REMPI) although this is the same process as RIMS. The most common multiphoton process for analytical purposes using molecules is resonant two photon ionisation (R2PI). In this process one photon excites a molecule to an excited electronic state and a second photon ionises the molecule. Since molecules normally have ionisation potentials between 7 and 13eV, R2PI requires UV photons. For many organic molecules the R2PI spectra are unique fingerprints but for others the spectra are broad and structureless at room temperature because of the complexity of the rotational and vibrational states. To improve selectivity these can be 'cooled down' to a few sharp peaks by introducing the molecules into supersonic beams (Tembreull et al. 1986).

Supersonic beam spectroscopy is normally carried out by seeding the molecules into a light carrier gas, eg Helium or Argon, at a few atmospheres pressure and expanding the gas through a narrow orifice into a vacuum. Rapid cooling down to a few degrees Kelvin is effected by converting the energy of the internal degrees of freedom into translational energy of the carrier gas via collisions. The apparatus for recording jet spectra by laser desorbing into a jet followed by laser ionisation is shown in fig. 11 (Grotemeyer and Schlag 1989).

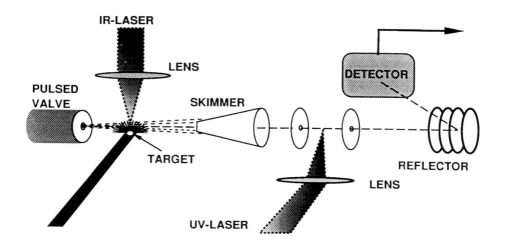

Fig. 11   The sample is irradiated by an IR laser to produce molecules which are seeded into a supersonic jet. These are post ionised by a UV laser and then analysed by a reflectron TOF mass spectrometer.

Fig. 12 (Marshall et al. 1990) shows a comparison of the room temperature R2PI spectrum of aniline with the single photon absorption spectrum and the jet cooled LIF and R4PI spectrum. Clearly the cooled spectra with its small number of sharp peaks is more likely to uniquely identify a molecule in a mass spectrometer. Low temperature wavelength spectroscopy using supersonic jet expansions has been used to distinguish between molecular isomers which is a very difficult mass spectroscopy problem (Blease et al. 1986).

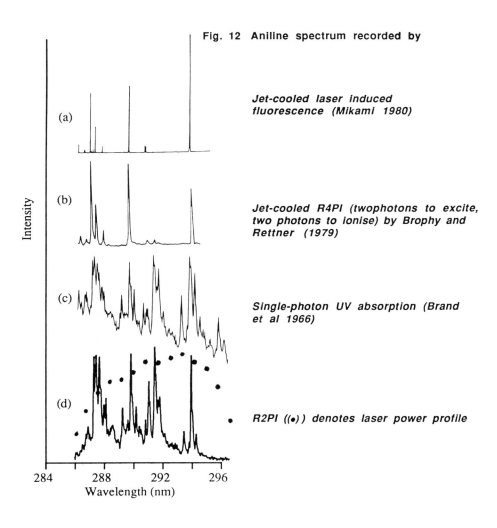

Fig. 12 Aniline spectrum recorded by

(a) *Jet-cooled laser induced fluorescence (Mikami 1980)*

(b) *Jet-cooled R4PI (twophotons to excite, two photons to ionise) by Brophy and Rettner (1979)*

(c) *Single-photon UV absorption (Brand et al 1966)*

(d) *R2PI ((•)) denotes laser power profile*

7.3 Observations on Carbon Clusters in a RIMS System

Although the following section is not strictly resonant ionisation mass spectroscopy it is included to illustrate related experiments which can be carried out using RIMS instrumentation.

A Nd-YAG laser was used to ablate a graphite target. The power was $10^8 Wcm^{-2}$, the pulse duration was 10ns and the spot size on the target was $1mm^2$. Carbon clusters, Cn with n up to 214 were observed in the reflectron (TOF) mass spectrometer. These are shown in fig. 13 and display all the prominent features observed by other authors (Rohlfing et al. 1984 and O'Keefe et al. 1986). Clusters of low n values are highly abundant. The intensity of observed clusters falls up to n = 31, beyond which a significant increase in their formation is measured. In addition, for n < 31, clusters for all values of n are obtained, but for n > 31 only those with even n values are observed. The clusters are clearly visible up to n = 214. For still larger n values there is an indication of clusters being present but the signal to background ratio is not good. It is interesting to note that between n = 32 and n = 214, the general envelope of cluster

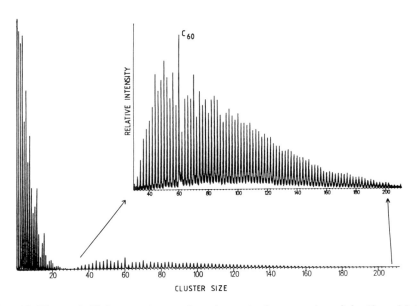

Fig. 13 Time of flight spectrum of carbon clusters produced in the ablation of a graphite target 532nm using a laser beam of $10^8 Wcm^{-2}$ flux.

formation and survival probability has a characteristic shape with a maximum around n = 50 and nearly linear decrease beyond that. Superimposed on the smooth envelope are clusters with n = 60 and 70 which are more abundant while n = 62 and 72 exhibit greatly depleted strengths.

Clusters with n = 60 (Buckminsterfullerene) and 70 have been shown theoretically to be particularly stable and are thought to have edgeless, spheroidal cage structures while the smaller clusters have linear chain or ring structures. The details of this experiment have been described in depth by Singhal et al. (1990) and it is argued that the different clusters may be formed with varying plume densities although it is pointed out that much detailed work with varying ablation laser power is necessary before a firm conclusion can be reached.

7.4 Resonant Laser Ablation (RLA)

It has been shown that laser ablation can be used to vaporise target materials for analytical purposes. This is the principle on which some analytical systems, e.g. laser microprobe mass analyses (LIMA and LAMMA), are based. The laser used in these systems is normally a quadrupled Nd:YAG at 266nm. A more sensitive but more complicated system is laser ablation followed by resonant post ablation ionisation of the neutrals which is a variant of RIMS requiring at least two laser systems. It has been felt that perhaps a simpler arrangement which might come close to the RIMS sensitivity would be Resonant Laser Ablation (McLean et al. 1990).

In this procedure a single tunable laser both ablates and resonantly ionises the sample if the laser wavelength is chosen to match known resonant atomic transitions. Verdun et al. (1987) first reported that such a procedure resulted in enhancements of ion production of about five-fold for Cd and Mo although the widths of the resonances were also particularly broad, 0.4 - 0.7nm. For these enhancements however the ablation procedure was carried out normal to the sample and in transmission mode. In the RLA procedure adopted by McLean et al., the mass spectrometer system shown in fig. 14 was used with the output from a tunable dye laser being used as the single resonant ablation laser. The beam was moderately focused to give a beam spot of about 1mm diameter and was directed at grazing incidence to the target, a semiconductor layer of $Al_x Ga_{1-x}$ (x = 0.3). Typical pulse energies were 1mJ for the red light and 100µJ for the doubled light. The ion signal for Al as a function of the ablation laser wavelength is shown in fig. 15. The resonance effects are clearly seen with a signal enhancement of greater than two orders of magnitude. The resonances are less than 0.05nm wide at half maximum height in sharp contrast with the work of Verdun et al. (1987). Careful calibration of the laser wavelength using a uranium hollow cathode lamp indicated that the resonances correspond to known atomic transitions.

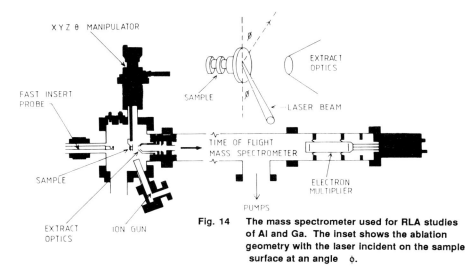

Fig. 14 The mass spectrometer used for RLA studies of Al and Ga. The inset shows the ablation geometry with the laser incident on the sample surface at an angle ϕ.

Recent work by Wang et al. (1990) has shown how the resonances behave as a function of grazing incidence angle as well as laser fluence. Furthermore it is suggested that the width of the transition which increases with both increasing laser fluence and angle of incidence is caused by atomic collisions within the plume.

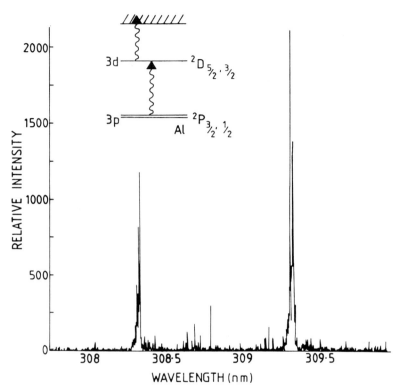

Fig. 15 The ion signal for Al as a function of the ablation laser wavelength. The enhanced yield at wavelengths corresponding to the excitation of the 3d from the ground state doublet is clearly observed.

CONCLUSIONS

RIMS has been shown to be a trace analysis technique of considerable potential both qualitatively and quantitatively. Already several commercial firms are advertising RIMS instruments of various types and in addition established laser microprobes can now be fitted with a post ablation ionisation capacity both resonant and non resonant. One of the difficulties which instrument manufacturers identify in the RIMS technology is the use of dye lasers with their inherent bulkiness and operating difficulties. With the development of solid state tunable lasers (Ti sapphire or optical parametric oscillators (Ledingham and Singhal 1990)) to replace the dye lasers it is likely that RIMS will soon reach its full potential and be accorded similar acceptance as SIMS

ACKNOWLEDGEMENTS

The author would like to acknowledge the considerable contribution of his colleagues and students in this research.

REFERENCES

Arlinghaus H F, Calaway W F, Young C E, Pellin M J, Gruen D M and Chase LL 1989 J. Appl. Phys. **65(1)** 281
Arlinghaus H F, Spaar M T and Thonnard N 1990 J. Vac. Sci. Technol. **A8(3)** 2318
Bekov G I and Letokhov V S 1983 Appl. Phys. **B30** 161
Bekov G I, Letokhov V S and Radaev V N 1985 J. Opt. Soc. Am. B/Vol. 2 1554
Benninghoven A, Rüdenauer F G and Werner H W 1987 *Secondary Ion Mass Spectrometry - Basic Concepts, Instrumental Aspects, Applications and Trends* (John Wiley & Sons, New York, Chichester, Brisbane, Toronto, Singapore)
Blease T G, Donovan R J, Langridge Smith P R R, Ridley T and Wilkinson J P T 1986 RIS 86, Inst. Phys. Conf. Ser.84 217
Brand J C D, Williams D R and Cook T J 1966 J. Mol. Spec. **20** 359
Brophy J H and Rettner C T 1979 Chem. Phys. Lett. **67** 351
Downey S W, Emerson A B and Kopf R F 1990 - *to be published in RIS 90, Varese, Italy, IOP Conference Series*
Fabbro R, Fabre E, Amiranoff F, Carban-Labaune C, Virmont J, Weinfield M and Max C E 1982 Phys. Rev. **A26** 2289
Fassett J D and Travis J C 1988 Spectrochim. Acta **43B** 1407
Fürstenau N 1981 Fresenius Z. Anal. Chem. **308** 201
Gladney E S 1980 Analytica Chimica Acta **118** 385
Goldsack T J, Kilkenny J D, MacGowan B J, Veats S A, Cunningham P F, Lewis C L S, Key M H, Rumsby P T and Toner W T 1982 Opt. Comm. **42** 55
Grotemeyer J and Schlag E W 1989 Acc. Chem. Res. **22** 399
Hurst G S, Nayfeh M Y and Young J P 1977 Phys. Rev. **A15** 2283
Hurst G S and Payne M G 1988a Spectrochim. Acta **43B** 715
Hurst G S and Payne M G 1988b *Principles and Applications of Resonance Ionization Spectroscopy* (Adam Hilger, Bristol England)
Kelly R and Dreyfus R W 1988 Nucl. Instr. and Meth. **B32** 341
Kimock F M, Baxter J P, Pappas D L, Kobrin P H and Winograd N 1984 Anal. Chem. **56** 2782
Ledingham K W D and Singhal R P 1990 - *Accepted for publication J. Anal. Atom. Spectr.*
Letokhov V S 1986 *Laser Analytical Spectrochemistry* (Adam Hilger, Bristol, England)
Lubman D M 1987 Anal. Chem. **59** 31A
Marshall A, Clark A, Jennings R, Ledingham K W D, McLean C J and Singhal R P 1990 - *to be published in RIS 90, Varese, IOP Conference Series*

Mamyrin B A, Karataev V I, Shmikk D V and Zagulin V A 1973 Sov. Phys. - JETP 37: 45

McLean C J, Marsh J H, Land A P, Clark A, Jennings R, Ledingham K W D, McCombes P T, Marshall A, Singhal R P and Towrie M 1990 Int. J. Mass. Spec. and Ion Proc. **96** R1

Mikami N, Hiraya A, Fujiwara I and Ito M 1980 Chem. Phys. Lett. **74** 531

Moore C E 1948 *Atomic Energy Levels* NBS Circular 467, US Government Printing Office, Washington DC

Oechsner H 1973 Z. Phys. **261** 37

O'Keefe A, Ross M M and Baronavski A P 1986 Chem. Phys. Letts. **130** 17

Pellin M J, Young C E, Calaway W F and Gruen D M 1986 Nucl. Instr. Meth. **B13** 653

Rohlfing E A, Cox D M and Kaldor A 1984 J. Chem. Phys. **81** 3322

Singhal R P, Land A P, Ledingham K W D and Towrie M 1989 J. Anal. Atomic Spectroscopy **4** 599

Singhal R P, Drysdale S L T, Jennings R, Land A P, Ledingham K W D, McCombes P T and Towrie M 1990 - *To be published in RIS 90 Varese, Italy, IOP Conference Series*

Tembreull R, Sin C H, Pang H M and Lubman D M 1986 Optics News October, p.16

Townsend P D, Kelly J C, Hartley N E W 1976 *Ion Implantation, Sputtering and their Applications* (Academic Press London) p.122

Towrie M, Drysdale S L T, Jennings R, Land A P, Ledingham K W D, McCombes P T, Singhal R P, Smyth M H C and McLean C J 1990 Int. J. of Mass Spec. and Ion Proc. **26** 309

Verdun F R, Krier G and Muller J F 1987 Anal. Chem. **59** 1383

Wang L, Borthwick I S, Jennings R, McCombes P T, Ledingham K W D, Singhal R P and McLean C J 1990 - *To be published in RIS 90 Varese, Italy IOP Conference Series*

A spectroscopic study of aniline using resonance ionization mass spectrometry

A Clark, A Marshall, R Jennings, K W D Ledingham, R P Singhal.

Department of Physics & Astronomy,
University of Glasgow,
Glasgow G12 8QQ.

ABSTRACT: Resonant two-photon ionisation mass spectrometry (R2PI/MS) is used to investigate the electronic structure of aniline vapour and its associated laser produced fragments in the region of the $^1B_2 \leftarrow {}^1A_1$ absorption band. All fragment ions so produced have the same wavelength dependence as the parent ion. An increase in detection efficiency may therefore be achieved by monitoring the total ion yield (parent + fragments) as a function of laser wavelength.

1. INTRODUCTION

Resonant two-photon ionisation (R2PI) of molecules coupled with mass spectrometric detection of the laser produced ions is an analytical technique which achieves both high sensitivity and selectivity (Boesl et al (1981), Schlag et al (1983)). Absorption of a laser photon, tuned to be in resonance with an allowed electronic transition of a particular molecule, excites that molecule to a bound excited state. Subsequent absorption of another photon from the same laser pulse achieves ionisation, provided that the energy of the intermediate state is greater than half of the first ionisation energy. Further photon absorption in the continuum can result in extensive fragmentation, yielding characteristic molecular fragmentation patterns.

2. EXPERIMENTAL.

A schematic of the experimental arrangement is shown in Fig.1.

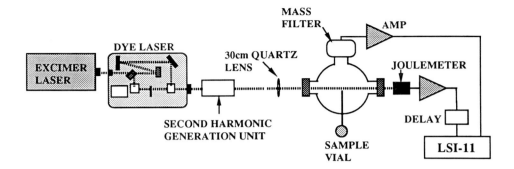

Figure 1: Experimental arrangement

A Lumonics XeCl excimer laser (model TE-860M) was used to pump a Lumonics EPD-330 dye laser. In this work, Rhodamine 6G laser dye with a tuning range of 570-595nm was used. The dye laser output was frequency doubled using KDP R6G second harmonic generation crystals, giving tunable UV laser radiation in the range 285-298nm. An Inrad autotracking system was employed to facilitate wavelength tuning through the required wavelength range.

Sample vapour was continuously admitted to the chamber through a 0.25mm diameter steel capillary tube controlled by a needle valve. The vapour was intercepted by the laser beam in the positively biased cage region of a quadrupole mass spectrometer (VSW Ltd). A schematic diagram of the laser-sample interaction region is shown in Figure 2 opposite.

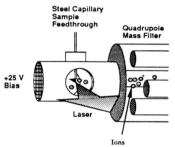

Figure 2. Laser/sample interaction region.

Laser produced ions/fragments were deflected into the spectrometer by a +25V bias on a wire grid, mass selected and detected by a Galileo 4771 Channeltron multiplier.

The electron signal produced was amplified by a Tennelec TC 205A linear amplifier. Typical flight times for ions were 20-40 μs, depending on their mass. A NIM gate pulse was generated at the correct time to coincide with both the ion signal and a delayed laser pulse energy signal from a Molectron J3-05 joulemeter. Both signals were digitised by the ADC of a LSI-11 data acquisition system. The timing electronics were triggered by a laser induced photodiode pulse.

3. RESULTS.

With the mass spectrometer tuned to transmit aniline parent ions, the resultant ion signal was recorded as a function of laser wavelength. The laser beam was focused into the interaction region by a 30cm quartz lens. A typical spectrum (laser fluence 10 MW/cm^2) is shown in Fig.3, along with the variation of dye laser power with wavelength. Four equally intense peaks are observed indicating possible saturation of the R2PI process at these wavelengths. Also shown in Fig. 3 is a low resolution single photon UV absorption spectrum recorded by Brand *et al* (1966), which also appears to saturate at these wavelengths.

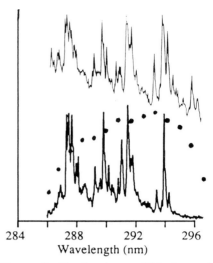

Figure 3. Single photon UV absorption spectrum (upper trace) & R2PI spectrum (lower trace) ● denotes laser power profile

A typical laser induced fragmentation pattern is shown in Fig. 4. The laser wavelength was tuned to the band origin transition and a complete mass spectrum was recorded.

The mass filter was tuned in turn to transmit each fragment ion of aniline and the laser wavelength was varied. Figure 5 shows the variation of fragment ion yield with wavelength for masses 77, 65 & 39. The fragment ion yield closely follows that of the

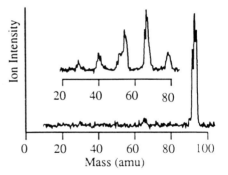

Figure 4. Mass spectrum of aniline showing fragments at higher gain.

parent, a result which arises because for any degree of fragmentation to occur, the parent molecule must absorb photons to produce excited states which may either pre-dissociate or absorb further photons. Fisanick *et al* (1980) with acetaldehyde and Zandee and Bernstein (1979) with benzene also reported strong similarities between parent and fragment ion production.

4. CONCLUSIONS

The observation that fragment ion production follows the same wavelength dependence as the aniline parent ions means that an increase in detection sensitivity for trace analysis work may possibly be realised by monitoring total ion yield as a function of laser wavelength instead of selecting only the parent ion, for example. This fact will also become important for detection of molecules that produce very small parent ion yields, e.g. the amino acids, PTH-Lysine & PTH-Cysteic acid (Engelke et al (1987)), and also for molecules such as nitrobenzene which may not produce any parent molecular ions.(Apel et al (1986)).

5. ACKNOWLEDGEMENTS.

This work was supported by the Ministry of Defence. A.Clark thanks S.E.R.C for a postgraduate studentship.

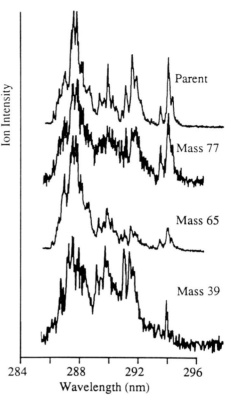

Figure 5. Variation of parent & fragment ion yield with wavelength.

6. REFERENCES.

E C Apel and N S Nogar 1986 *Int. J. Mass Spec. & Ion Proc.* **70** 243
U Boesl, H J Neusser and E W Schlag 1981 *Chem. Phys.* **55** 193
J C D Brand, D R Williams and T J Cook 1966 *J. Mol. Spec.* **20** 359
F Engelke, J H Hahn, W Henke and R N Zare 1987 *Anal. Chem.* **59** 909
G J Fisanick, T S Eichelberger, B A Heath and M B Robin 1980 *J. Chem. Phys.* **72** 5571
E W Schlag and H J Neusser 1983 *Acc. Chem. Res.* **16** 355
L Zandee and R B Bernstein 1979 *J. Chem. Phys.* **71** 1359

Inst. Phys. Conf. Ser. No. 113: Section 5
Paper presented at International Meeting on Optogalvanic Spectroscopy, Glasgow, 1990

Resonance ionization mass spectrometry applied to the trace analysis of gold

P T McCombes, I S Borthwick, R Jennings, K W D Ledingham and R P Singhal

Department of Physics and Astronomy, University of Glasgow, Glasgow G12 8QQ.

ABSTRACT: Resonance Ionisation Mass Spectrometry with laser ablation for sample vapourisation has been used to analyse trace amounts of gold in a copper matrix. Ionisation is achieved using a simple two photon scheme. Good linearity was seen between the gold concentration in the sample and the ion signal. The detection limit determined by these experiments was ~200ppb. Comment is made on how this sensitivity could be increased by at least an order of magnitude.

1. INTRODUCTION

Although Resonance Ionisation Mass Spectrometry (RIMS) has been shown to be of great use in research laboratories [Fassett and Travis (1988)] a ruggedised RIMS instrument had not, until recently, been built commercially. It was for this purpose that an instrument was designed at Glasgow and built by Kratos to carry out routine analysis with essentially unprepared samples [Towrie *et al* (1990)]. This instrument has demonstrated trace analysis at the part per million (ppm) sensitivity using NIST coal samples with a large carbon content. The formation of carbon clusters essentially defined the attainable sensitivity [Ledingham *et al* (1990)]

In order to determine the instrument's sensitivity and dynamic range linearity a series of copper samples containing well characterised amounts of gold were examined. It is obviously of great importance that an instrument can measure both major and minor traces. Furthermore the detection of precious metals, in particular at ppm sensitivity, is of considerable commercial interest.

© 1991 IOP Publishing Ltd

2. EXPERIMENTAL ARRANGEMENT

Figure 1 is a schematic diagram of the Glasgow Resonance Ionisation Mass Spectrometer. A detailed description of the apparatus has been given by Towrie *et al* (1990) and only the relevant details are given here.

The samples were each mounted on a stainless steel sample stub using non-conducting epoxy. Conducting epoxy, which is normally used for this purpose, is unsuitable as it contains silver, which is likely to have gold as an impurity.

The sample stub is mounted on a $XYZ\theta$ precision manipulator at the centre of the spherical sample chamber. The system is maintained at a pressure of 10^{-9} Torr by turbomolecular and diffusion pumps. Sample changeover is possible in ~10 minutes using a rapid transfer probe.

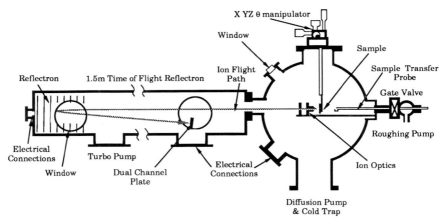

Fig. 1. The Glasgow Resonance Ionisation Mass Spectrometer

The sample is vapourised by a Quantel Nd:YAG 585 laser with harmonic generating crystals. The pulse is ~10ns in duration and has a nearly Gaussian spatial profile. The laser is focused onto the sample by a lens (f=30cm) to give a spot size on the sample surface of ~1mm^2. The angle between the ablating beam and the sample surface is 45^0.

The post ablation ionising laser system consists of a Spectron Nd:YAG laser with frequency doubling, tripling and quadrupling facilities which pumps two dye lasers with wavelength ranges of 540-700nm and 400-700nm. The wavelength range can be extended down to 220nm by passing the dye laser output through a frequency doubling crystal mounted on an Inrad autotracker.

The ionising laser is lightly focused to a waist of ~1mm by a quartz lens and passes approximately 2mm above the sample surface. The timing between the ablation laser and the ionising laser is adjusted so that the ionising laser interrogates the largest possible number of neutrals.

The ions are accelerated into the time of flight region by a positive voltage of 2-3kV on the sample stub. High acceptance angle optics and x-y deflector plates combine to give high ion transmission. The reflectron time-of-flight system provides energy focusing and reduces the background due to ions created by the ablation laser. A thin wire follows the ion flight path and acts as an electrostatic ion guide. The ions are detected by a Galileo multichannel plate detector.

The signal from the detector is digitised by a LeCroy 2261 transient recorder which provides 640 time channels 10ns long, each of 11 bit resolution, giving a window on the data of 6.4µs. A COMPAQ 386/25 is used to analyse the data and control the data acquisition process. Timing signals for the equipment are provided by a Stanford DG535 digital delay generator which provides the timing of the transient data window and also the delay for the ADC gates. The delay between the lasers is controlled by a custom built digital delay generator.

3. RESULTS AND DISCUSSION

The resonance ionisation scheme used in this paper is absorption of one photon at 267.67nm from the ground state ($5d^{10}\,6s\;^2S_{1/2}$) to the $5d^{10}\,6p\;^2P^o_{1/2}$ excited state followed by ionisation after the absorption of a second photon at this wavelength. This scheme has been used previously by Krönert *et al* (1987). The high quality samples used consisted of equal parts of gold and silver homogeneously distributed through a copper matrix. The concentrations of gold ranged from 1500ppm to 10ppm.

The wavelength used for ablation was 532nm. The optimal ablation power was determined by taking several spectra at different powers on the 1500ppm sample and finding the best signal to noise ratio. This corresponded to a fluence of about 2mJ/mm^2. Figure 2 shows the effect of changing the delay between the lasers on the gold ion signal. Spectra in this case were averaged over 300 shots to eliminate as much as possible the variations in signal size caused by random fluctuations in the laser output powers. The optimal time delay between the lasers was determined to be 1.8µs.

The sensitivity of the method is shown in Figure 3, the signal was accumulated over 10000 shots using a lightly focused laser beam (0.4mm beam diameter) with a pulse energy of

100μJ. This is sufficient power to saturate the bound-bound transition, but falls far short of saturating the ionisation step. The signal to noise ratio is ~50:1, indicating an ultimate detection level for the arrangement used of ~200ppb. The analysis time was 15mins. Figure 4 shows a series of spectra taken from 1000 shots under the same conditions on the 100ppm sample, this illustrates the reproducibility of the system.

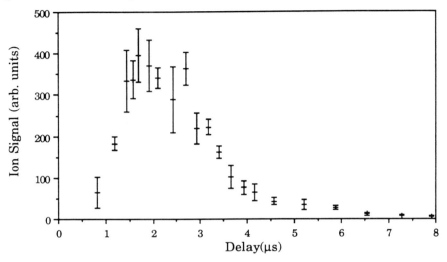

Fig. 2. Temporal distribution of gold atoms ablated by 532nm Nd:YAG. The data was averaged from 3 runs and the error bars show the standard deviation.

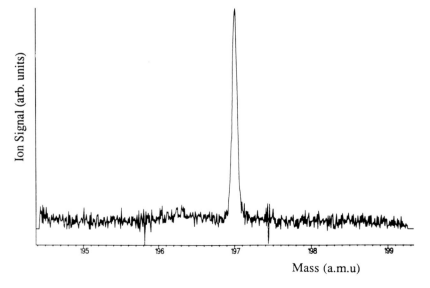

Fig. 3. RIMS signal for gold at 10ppm level accumulated over 10000 shots.

Another important aspect of this work is the linearity of the gold signal. Figure 5 shows a graph of gold concentration against the signal size. This data was taken for 1000 shots. The graph has a gradient of 1.055, indicating a strong degree of linearity of the signal.

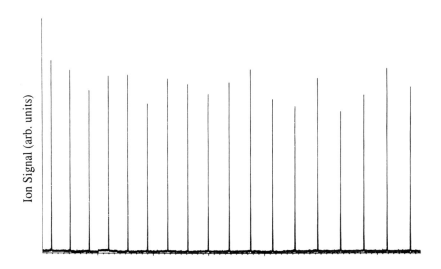

Fig. 4. A series of gold peaks from 100ppm sample each accumulated over 1000 shots. The standard deviation of the integrated peak heights is 10%. This illustrates the reproducibility of the RIMS signal.

4. CONCLUSIONS

A detection level of ~200ppb was calculated for the resonant scheme used. By using the 266nm output from the Nd:YAG pump laser saturation of the ionisation step should be possible [Lee *et al* (1987)] resulting in a detection level of 10-20 ppb.

In addition the efficiency of the detector was impaired due to age, and it is estimated that a new detector would lower the detection limit further to ~5ppb.

5. ACKNOWLEDGEMENTS

It is a pleasure to thank Stephen Chryssoulis of Surface Science Western, Canada for the use of the gold samples. P T McC and I S B would like to thank the SERC for studentships.

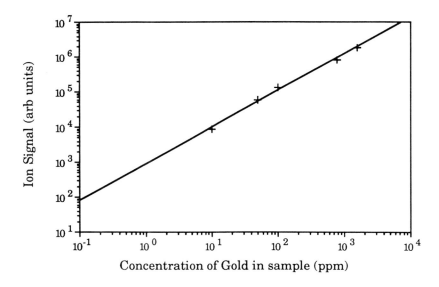

Fig. 5. Gold concentration against RIMS signal illustrating the linearity of the process

6. REFERENCES

Fassett J D and Travis J C 1988 Spectrochim. Acta. 43B pp1408-1422

Krönert U, Becker St, Hilberath Th, Kluge J and Schulz C 1987 Appl. Phys. A44 pp339-345

Ledingham K W D, Singhal R P, Drysdale S L T, Jennings R, Land A P, McCombes P T and Towrie M 1990 To be published in Resonance Ionisation Spectroscopy 1990, IOP Conference Series

Lee J K P, Raut V, Savard G and Thekkadath G 1987 Resonance Ionisation Spectroscopy 1986, IOP Conference Series No 84, pp139-144

Towrie M, Drysdale S L T, Jennings R, Land A P, Ledingham K W D, McCombes P T, Singhal R P, Smyth M H C and McLean C J 1990 Int. J. Mass Spectrom. Ion Proc. 96 pp309-320

New directions in high resolution optogalvanic spectroscopy

Antonio Sasso

Dipartimento di Scienze Fisiche, Università di Napoli
Mostra d'Oltremare Pad 20, 80125 Napoli, Italy

ABSTRACT: A review of the principal features of the optogalvanic technique is presented for a variety of spectroscopic investigations. The advantages and limits are examined in some detail for specific cases of atomic species produced in different discharge configurations (electrodeless discharges, hollow cathode, positive column). Alternative approaches to improve sensitivity and resolution are discussed.

1. INTRODUCTION

The development of new spectroscopic techniques has brought a deeper knowledge of atomic and molecular structure. In particular the advent of laser sources has made it possible to revive "old" spectroscopic techniques and to realize completely "new" ones (Ernst and Inguscio, 1988). Optogalvanic (OG) spectroscopy, at least its simpler version, belongs to the former category. In fact, the OG effect was first observed by Penning in 1928 but its application to spectroscopy dates from the availability of tunable lasers (Green *et al*, 1976). The optogalvanic effect appears in gas discharges illuminated by radiation resonant with an atomic or molecular transition of the species within the discharge. Usually OG detection is based on changes in the discharge current or, equivalently, variations in the discharge impedance. A very simple description of the OG effect is given in terms of the difference between the ionization cross−section for the two atomic or molecular levels whose populations are perturbed by resonant radiation. Nevertheless the numerous and complex processes involved in a discharge prevent us from developing a general quantitative theory (Doughty and Lawler, 1983, Sasso *et al* 1988a, De Marinis *et al* 1988).

© 1991 IOP Publishing Ltd

Optogalvanic spectroscopy has become a very popular technique thanks to many features that make it very attractive in comparison to other conventional spectroscopic techniques. First it is an easy and low cost technique since its detection scheme is based on very simple circuits and it does not require the use of expensive devices, such as monochromators, phototubes, etc.

OG spectroscopy is intrinsically more sensitive than absorption and fluorescence. In fact, OG detection is based upon a signal on a zero background while the detection of weak absorption or resonant fluorescence is sometimes problematic due to the superposition of a large background.

The high sensitivity of OG spectroscopy derives from the high efficiency of integral charge collection. From this point of view OG spectroscopy is similar to resonance ionization spectroscopy (Hurst *et al*, 1979) where atoms or molecules are ionized through the absorption of two or more resonant photons and the produced electron-ion pairs are detected by means of electrodes.

OG spectroscopy is very versatile because atoms, molecules, ions, and radicals can be investigated over a wide spectral range (for reviews see, for example, Ernst and Inguscio, 1988, Ochkin *et al*, 1986, Barbieri *et al*, 1990 and all the papers appearing in Journ. de Phys. Coll. No 7, tome 44, 1983).

OG spectroscopy is particularly advantageous for the study of highly excited levels that are produced in a discharge and in molecular discharges it is the natural way to investigate atomic, molecular or ionic species available from the parent molecular compound.

Moreover, vapours of refractory elements are accessible to OG spectroscopic investigation because they are easily produced in hollow cathode discharges by sputtering.

All these considerations show the usefulness of the OG technique to a wide range of applications, such as wavelength calibration, laser frequency stabilization, detection of trace elements in flames, etc.

Furthermore a lot of sub-Doppler techniques have been successfully adapted to OG detection, such as intermodulated OG spectroscopy (IMOGS) (Lawler et al, 1979), polarization intermodulated excitation spectroscopy (POLINEX) (Hänsch et al, 1981), and two-photon spectroscopy (Goldsmith et al, 1979). The success of these high-resolution techniques is not so obvious if we keep in mind the complexity of the plasma medium in which OG detection is performed.

Nevertheless some disadvantages limit the OG technique when high resolution and sensitivity are required. With regard to the signal-to-noise ratio, the current shot noise limit (represented by the statistical fluctuation of the current) is achieved only when the discharge is driven in a narrow range of discharge parameters. These conditions are not fulfilled at high currents which is necessary, for instance, for efficient sputtering in hollow cathode lamps. The OG line profile may be sensitively different from profiles obtained by absorption spectroscopy (Kane, 1984). This result is not surprising if we consider the complex mechanism leading to the OG signal. In addition,the presence of strong electric fields occurring in a particular discharge configuration (cathode region) can significantly affect the homogeneous line profiles.

OG detection can be problematic when UV transitions are investigated by means of very low cw laser power (hundreds of mW). This is still more evidenced for transitions starting from the ground state because, in these cases, the fractional changes in discharge impedance can be negligible.

In the case of sub-Doppler techniques, such as intermodulated optogalvanic (IMOG) spectroscopy, the lineshapes are affected by large pedestals due to velocity-changing collisions. If this pedestal can provide useful information in collisional processes acting in the discharge (Sasso et al 1988b) it represents a serious limit for resolution. The Doppler-background can be removed by modulating the laser intensity with frequency higher than the collisional frequency. This offers the further advantage of reducing the 1/f laser amplitude noise. Nevertheless, when high frequency devices are to be used, other techniques are much more convenient in comparison to the OG scheme. For example, Hollberg et al (1983) have shown that with highly sophisticated Fast Modulation spectroscopy the signal-to-noise ratio could be enhanced by three orders of magnitude with respect to OG spectroscopy. In order to improve resolution and sensitivity, our group has explored other techniques, where the observable is not the population but the polarization. Polarization spectroscopy (Wieman and Hänsch, 1976) represents a well known Doppler-free technique suitable to investigate atoms or molecules "prepared" in glow discharges.

In this paper we report a review of the more recent activities performed by our group at the University of Naples in the field of the OG spectroscopy. Many of the advantages and limits of OG spectroscopy briefly presented above will be discussed in some detail for specific spectroscopic investigations of atoms.

In Section 2 we report on high resolution spectroscopy of atomic oxygen produced in a radio frequency discharge. The spectral features of this element are of fundamental importance because accurate spectroscopic investigations allow us to test quantum mechanical calculations still possible for a so light and hence tractable element. Moreover isotope shift and hyperfine structure measurements provide precious information in nuclear physics because the nucleus of oxygen is doubly magic. Refractory elements represent another interesting case of species advantageously "prepared" in a discharge. These elements are of remarkable interest for astrophysics, standards of frequency and isotope shifts. Their resonances are in the UV region, only recently accessible to high spectral purity laser sources. Nevertheless, as will be shown in Section 3, due to the low intensity available, the investigation of these elements requires alternative techniques to the optogalvanic scheme.

In Section 4 we present a new approach to "impedance" spectroscopy where the physical parameter monitored is not the discharge current but the dielectric constant. The sensitivity achieved with this technique is an order of magnitude higher than "conventional" optogalvanic detection and makes this technique promising for analytical spectroscopy in flames.

2 OPTOGALVANIC AND POLARIZATION SPECTROSCOPY OF ATOMIC OXYGEN

2.1 Atomic production: OG monitoring of O_2-dissociation

The difficulties for spectroscopic investigation of atomic oxygen arise with the production of excited atoms since no allowed optical transitions are available from the 2^3P ground state. Recently we made oxygen accessible to high resolution spectroscopy (Inguscio et al 1988, Sasso et al 1988c). However, on the way to carrying out spectroscopic investigations, we encountered a number of interesting phenomena related to the production of the atom and its detection. Oxygen atoms were produced by dissociation of molecular oxygen in a radio frequency discharge (power $\approx$ 50 Watt, frequency $\approx$ 60 MHz). The discharge was maintained by an oscillator fed by a current stabilized power supply (Fig 1). The oxygen atom density was monitored by means of fluorescence and OG techniques. In particular OG signals were detected by monitoring the feed-back loop between the oscillator and the current stabilized power supply.

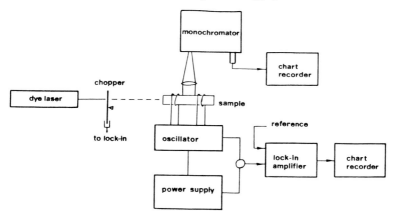

Figure 1. Atomic oxygen is produced through the dissociation of O_2 in an electrodeless discharge sustained by a noble gas. Atoms are monitored by fluorescence and optogalvanic detection.

Figure 2 shows typical fluorescence and OG spectra around the oxygen multiplet $3p^5P_{1,2,3}-4d^5D_{4-0}$ at 616nm obtained in a Ne-O_2 r.f. discharge. The resolution offered by OG detection, in spite of the broad band operation of the dye laser, allows us to resolve the fine structure of the lower level $3p^5P_{123}$ (Δ_{12} = 59.3 GHz, Δ_{23} = 110.3 GHz). In the spectra of Fig 2 two strong neon transitions starting from metastable levels are also recorded. The presence of neon in a discharge with a low density of molecular oxygen (Ne/O_2 $\approx$ 10/1) produces very efficient formation of oxygen atoms.

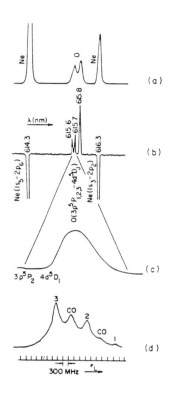

Metastable noble gas atoms are produced by electron impact excitation with large cross sections and collisions with O_2 molecules produce excitation, dissociation, or ionization of the target molecules.

The different energies of metastable rare gases with respect to the O_2 potential curves (see Fig 3) originate different dissociative pathways. From this point of view noble gases can be lumped in two groups. The first one concerns atoms such as neon and helium that give rise to Penning ionization; dissociative recombination of the molecular ions produced leads to oxygen atoms in high excited states:

Figure 2 Fluorecence (a) and OG (b) spectra obtained in a O_2–Ne rf discharge. The opposite OG sign for neon and oxygen lines denotes the different mechanisms leading to the OG signal. The recording in (b) was obtained by frequency scanning a multimode dye laser. The Doppler–limited (c) and IMOG (d) recording was obtained using a single–mode dye laser.

In the case of Ar, Kr, and Xe, oxygen atom formation seems to occur through the formation of a quasi–molecule followed by dissociation:

$$Ne^* + O_2 \dashrightarrow Ne + O_2^+ + e^-$$
$$He^* \qquad\qquad He \qquad\qquad (1)$$

$$O_2^+ + e^- \dashrightarrow O + O^*$$

$$\text{Ar}^* + \text{O}_2 \dashrightarrow \text{Ar} + \text{O}^*(2^1\text{D}, 2^1\text{S}) + \text{O}(2^3\text{P}) + \Delta E \quad (2)$$

The extra energy ΔE in Eq (2) is released as translational energy of the reaction products and hence, is responsible for a wider velocity distribution ("warm" atoms in non-thermal equilibrium).

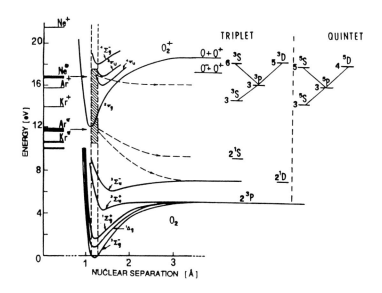

Figure 3   Relevant atomic levels and molecular potential curves of O, O_2, and O_2^+ with the relative energies of some metastable rare gases used in this experiment.

Collisions with electrons can transfer this kinetic energy excess to higher excited states and, as a consequence, the Doppler broadening of the lines can be significantly altered.   This phenomenon, first observed by Feld et al (1973) in the study of oxygen laser transitions, is illustrated in Fig 4 which shows the Doppler-limited lineshapes of transitions involving triplet and quintet levels obtained in Ne-O_2 and Ar-O_2 mixtures.   Wide Doppler broadening is observed only when Ar is used as buffer gas and only in the spectra of triplet transitions (Fig 4d).

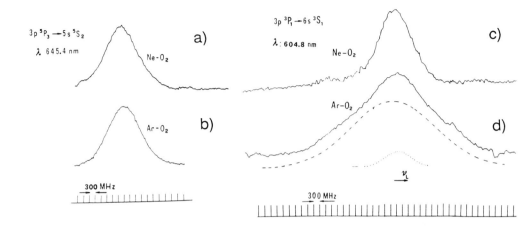

Figure 4 Doppler-limited OG line profiles of the $3^5P_3-5^5S_2$ (λ=645.4nm) (a,c) and $3^3P_1-5^3S_1$ (λ=609.6nm) (b,d) oxygen transitions obtained by using Ar and Ne as buffer gas. The "anomalous" broadened lineshape is fitted by two Gaussian curves following a model described in the text.

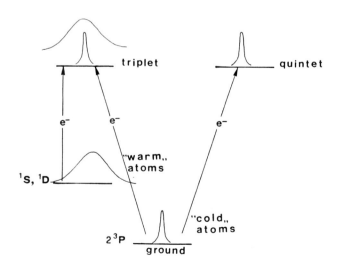

Figure 5 Schematic diagram of electron impact excitation pathways which are regulated by spin-conservative selection rules.

These Doppler "anomalies" can be explained by means of the model depicted in Fig 5 (Sasso *et al* 1990). The excitation of low energy atoms by electron impact is regulated by spin-dependent selection rules. In particular singlet "warm" atoms can be excited into triplet or singlet configurations while thermalized ground state atoms may be excited into triplet or quintet states. As a result, atoms in highly excited triplet states exhibit a velocity distribution given by the superposition of two Gaussians. This phenomenological model allows us to explain the anomalous Doppler broadened lineshape of Fig 4d. Indeed, a good fit of this line profile has been obtained in this way. The width of the narrower Doppler Gaussian is about 2 GHz and corresponds to a temperature ≈ 500K while the broader one corresponds to a temperature ≈ 4500K, i.e. nearly one order of magnitude larger.

Atomic oxygen so prepared could be used to explain interesting phenomena observed in oxygen-enrichment of high Tc $Y_1Ba_2Cu_3O_{7-x}$ superconductive films. Indeed, recently, the oxidation process has been demonstrated to be particularly efficient when performed with a reactive atomic oxygen source generated from a microwave discharge.

2.2 Fine Structure

Accurate measurements of fine structure in atomic oxygen are of interest because oxygen is relatively light and hence theoretically still tractable. Champeau *et al* (1975) have demonstrated, by *ab initio* calculations, that the various magnetic interactions, spin-orbit, spin-other-orbit, spin-spin are relevant if higher-order terms are taken into account. The accuracy of fine structure data reported by Moore (1976) by conventional spectroscopic techniques was limited by the Doppler effect. This made the determination of fine structure splittings comparable with the Doppler width still more difficult. Recently we succeeded in performing the first Doppler-free investigations of optical transitions by using intermodulated optogalvanic (IMOG) spectroscopy. Fig 2d shows the IMOG spectra of the 3 $^5P_2-4^5D_{3,2,1}$ oxygen transition: three fine-structure components are resolved, together with the associated crossover peaks originating from two transitions sharing the lower level and overlapped within the Doppler width. Fine structure splittings are determined with an accuracy one order of magnitude higher than with data from conventional spectroscopy. But the comparison with Doppler-limited data is also interesting for verifying some assumptions. For example, it can be checked whether the relative intensity of f.s. components follows the theoretical values in the Russell-Saunders coupling scheme (Sasso et al 1988c).

High resolution offered by IMOGS has also been the starting point of an interesting analysis of isotope shift of atomic oxygen as will be discussed in the next section.

2.3 Isotope Shifts

As is well known, the isotope shift (IS) in atomic transitions is due to the mass and volume (field) contributions. The mass effect is the result of a normal contribution $\Delta \nu_N$, which is easily computed and equivalent to that for a hydrogen-like system, and of a specific contribution (SM) which takes into account electron cross interactions. The field contributions (FS), originate from the change in the nuclear charge distribution. Both the SM and FS contributions are difficult to calculate and require accurate knowledge of electronic wavefunctions and nuclear parameters.

The FS is normally negligible for light elements (A < 30) but its observation is more likely to occur for "magic" number nuclei (nuclei with closed neutron and/or proton shells). The nucleus of ^{16}O is doubly magic, making the investigation of the isotope effect stimulating. In particular, an accurate determination of the FS is interesting in nuclear physics because this effect is proportional to the variation of the mean square nuclear-charge. Fig 6 shows an IMOG recording for the $3p\,^5P_3 - 5s\,^5S_2$ transition (λ=645.6nm) obtained with a ^{18}O enriched sample. The Doppler width for this transition ($\approx$2GHz) should prevent observation of any isotopic effect. Nevertheless FS cannot be evidenced from the IS of a single isotope pair even if several transitions are considered. It can be done using a semiempirical approach in which the residual modified shift (RMS) of several isotope pairs for two different transitions λ_i and λ_j are plotted against each other (King plot, King 1984). In the presence of FS the experimental points are distributed around a straight line whose slope gives the FS ratio for the two transitions. The occurrence of three stable isotopes ^{16}O, ^{17}O, ^{18}O is, therefore favourable since at least three isotopes are necessary to distinguish the volume effect from the predominant mass effect. By looking at the IMOG spectra of Fig 6 we can observe that the homogeneous lineshapes are affected by a large and broad pedestal due to

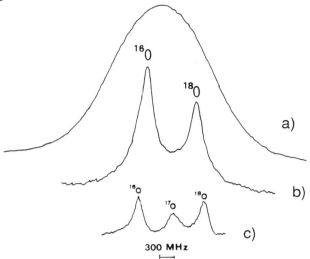

Figure 6 Doppler broadening (a) masks any isotopic effect. IMOGS recording (b) for the $3^5P_2-5^5S_2$ transition performed using an ^{18}O enriched sample. Polarization spectroscopy (c) extended to an $^{17}O-^{18}O$ enriched sample for the $3^5P_3-4^5D_4$ transition.

velocity-changing collisions (VCC) which should prevent the resolution of the line from the third oxygen isotope, ^{17}O. In order to improve sensitivity and resolution we have used polarization spectroscopy (Gianfrani et al 1990). The experimental apparatus is schematically shown in Fig 7 while a typical recording is shown in Fig 6c. It may be noted that the residual Doppler pedestal is now completely removed, allowing us to clearly resolve the three isotopes. The physical reason for this is the fact that orientation is not conserved during a VCC. A further advantage of polarization spectroscopy is represented by a simplificaiton of the spectra for transitions exhibiting fine structure splitting comparable with the IS as those shown in Fig 6c. Indeed, in that spectrum, only one of the three fine structure components, namely the $^5P_3-^5D_4$ is observed. This is due to the different relative fine structure intensities calculated from a sum of the Clebsh-Gordan coefficients over the m_j levels (see, for example, Demtroeder, 1988).

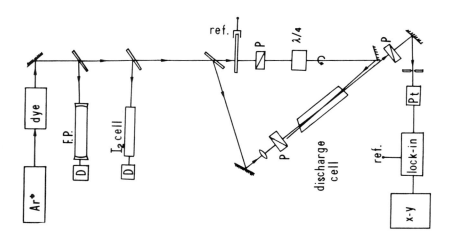

Figure 7 Experimental apparatus for polarization spectroscopy.

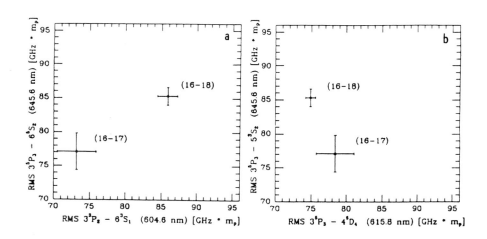

Fig 8 Analysis of the isotope shifts using King plots.

In Fig 8 the RMS in the $3^5P_3-5^5S_2$ transition is plotted against the RMS in the $3^3P_2-6^3S_1$ (Fig 8a) and $3^5P_3-4^5D_4$ (Fig 8b) transitions. Since the points for the two pairs of isotopes do not overlap, it means that a non-negligible FS effect is present. In addition, the slopes of the two King plots are consistent with the expected behaviour of the field effect. Indeed, for transitions involving S states the FS is more pronounced because these orbitals are more penetrating in the nucleus.

Furthermore, unlike ^{16}O and ^{18}O, whose nuclear spins are zero, ^{17}O has a nuclear spin (I=5/2). The unresolved hyperfine structure (hfs) is responsible for the broader linewidth of the ^{17}O peak in Fig 6c. In the next paragraph we will see, thanks to the recent development of new very low-noise semiconductor diode lasers, that it has been possible to successfully resolve (for the first time) the hfs of excited levels of atomic oxygen.

2.4 Hyperfine structure

As mentioned before, ^{17}O is the only stable oxygen isotope with a nuclear moment. The only previous hfs investigation was performed on the 2^3P ground state in a paramagnetic resonance experiment. Very recently we have extended hfs studies to higher excited states, investigating the transition between the 3^5S_2 and the $3^5P_{1,2,3}$, at 777nm. Several features make this transition extremely interesting. First, it starts from the lowest S level thus nuclear effects are expected to be particularly strong. The $^5S_2$ level is metastable, with a lifetime of 180µs, and is connected to the ^3P ground state by an intercombination line at 136nm. In addition, the $^5S-^5P$ transitions exhibit a strong oscillator strength and are accessible to the GaAℓAs/GaAs diode laser. Furthermore, radiative cooling of an atomic oxygen beam through this transition is the only realistic scheme to obtain "cold" atoms in the ground state, because no optically allowed transitions are possible from this level.

The experimental arrangement used to investigate this transition is described in detail in the paper by Tino *at al* (1990). The extremely low noise exhibited by the diode laser, combined with a high sensitivity detection scheme, allowed us to detect oxygen at very low pressure (a few mTorr).

Fig 9 shows a Doppler-free spectrum obtained using an enriched sample by means of saturated absorption spectroscopy. Besides the individual resonances of ^{16}O and ^{18}O, the hfs of the lower $^5S_2$ level is fully resolved for ^{17}O. In addition, the hfs of the upper $^5P_3$ level, unlike the $^5P_3-^5D_4$ transition shown in Fig 6c, is partially resolved in this case. The measured hyperfine magnetic constant A for the $^5S_2$ level is 74.3(4)MHz and from this it can be estimated that the electron density at the nucleus $|Y(O)|^2 = 0.12$ (in Bohr radius units). This data is of remarkable interest for an absolute estimation of the volume effect on the isotope shift.

The resonances of Fig 9 are near the radiative line width and allow us to give a lower limit to the lifetime of the $^5P_3$ level (t≈6ns). This lifetime plays an important role in the radiative cooling scheme mentioned before. In fact, the deceleration time depends on the number of radiative optical cycles and, for t=6ns and an initial velocity of 4.10^4cm/s, t_{decel}≈150 μs. This time is lower than the lifetime of the lower $^5P_3$ metastable level demonstrating the possibility of cooling a thermal oxygen beam. Cold oxygen atoms could be produced in the ground state through the decay of cooled $O(^5P_3)$-atoms and would be suitable for experiments of interest in metrology. Indeed, the strongly forbidden $^3P-^1D$ transition at 630nm could be accessible with an atomic sample interacting with electromagnetic radiation for a sufficiently long time.

3. HIGH RESOLUTION ULTRAVIOLET SPECTROSCOPY OF REFRACTORY ELEMENTS

Recent progress in nonlinear crystal technology have made it possible to generate ultraviolet (UV) cw radiation by frequency doubling tunable dye lasers. This makes many refractory elements accessible to Doppler-free spectroscopic investigations since they have transitions in the UV.

The production of refractory vapours normally requires ovens operating at high temperature. Nevertheless densities of such elements suitable for spectroscopic studies are easily produced in hollow cathode discharges by sputtering. In addition these discharges are efficient generators of ionic species.

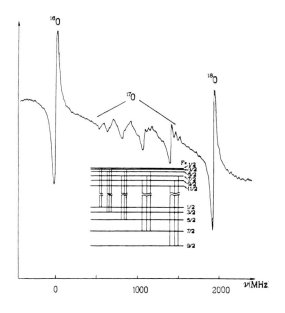

Figure 9 Saturated-absorption spectrum for $^3S_2 - {}^3P_3$ oxygen transition obtained using a diode laser. Hfs of $^5S_2$ level is fully resolved for ^{17}O.

In our laboratory we have started a study of several refractory atoms including titanium, calcium, magnesium, strontium, iridium, and osmium. These atoms are of interest for astrophysics because their abundances in spectroscopic observations give information on nucleo-synthesis processes occurring in stars. They are also good candidates as standards of frequency and time because neutrals and ions of these elements can be studied at rest by means of radiative cooling or ion traps. Several features of their spectra, such as isotope shifts, fine structure and hfs, are unknown or known only with the accuracy provided by conventional spectroscopy. For example, until now, little information has been available on the IS of Ti in spite of its large number of stable isotopes (46,47,48,49,50). The only IS measurements on titanium have been reported by Maruyama *et al* (1987) in an experiment investigating isotope separation. In that case, however, the accuracy was limited by the linewidth of the pulsed laser.

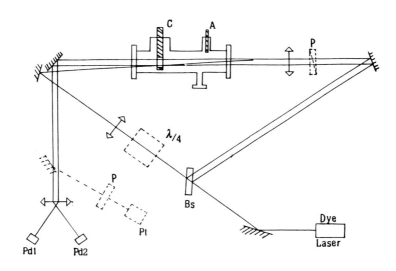

Figure 10 Experimental apparatus for saturated–absorption and polarization spectroscopy in the UV. Titanium atoms and ions are produced by sputtering in a hollow cathode discharge.

Recently we have performed the first Doppler–free spectroscopic investigation of TiI and TiII in order to provide accurate IS and fine structure splitting data. Our experimental apparatus was configured for saturated–absorption and polarization spectroscopy (see Fig 10). We also tried to monitor UV transitions by OG detection, but no signal was observed in this case. In fact, the low UV laser intensity available and the large discharge noise arising from the high current regime makes the use of OG detection problematic since the fractional changes of impedance are not detectable. On the contrary, OG monitoring was possible for the transition at 629nm arising from the metastable $4s\,^3F_4$ level, excited by the discharge. In saturation spectroscopy the noise due to the laser intensity fluctuations was reduced by balancing and subtracting the signals of two probe beams, only one of which was interacting with the pump beam.

In Fig 11 absorption and polarization spectroscopy are compared for the $4s^2$ $^3F_2-4p\,^3G_3^0$ TiI transition at $\lambda = 334.3$nm. As for IMOG spectroscopy the spectra obtained with saturation spectroscopy are affected by a residual Doppler pedestal due to VCC. In addition, the better signal-to-noise ratio of the polarization recording, allows us to distinguish the even isotopes (46,48,50) clearly and partially resolve the hfs of the odd isotopes (47,49).

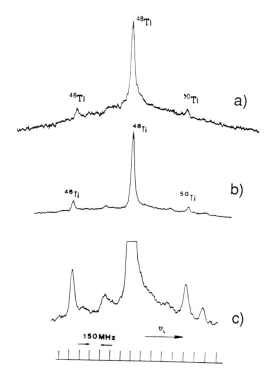

Figure 11 Saturated-absorption (a) and polarization (b) recordings of the TiI resonance transition at 334.3nm. In (c), the same spectrum obtained after a magnification of 4.5. The presence of additional peaks is due to the hfs of the odd ^{47}Ti and ^{49}Ti isotopes.

Isotope shifts have also been studied for several transitions of TiII and an analysis of these results is in progress.

4. A NEW APPROACH TO "IMPEDANCE" SPECTROSCOPY

Several detection schemes have been developed to monitor changes of impedance induced by resonant radiation. In most cases the discharge current is still the physical parameter which is observed.

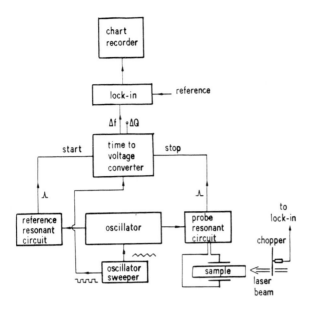

Figure 12 Experimental set-up for detection of changes in the dielectric constant. The sample, not necessarily a discharge, is part of the capacitance of the probe resonant circuit. Changes of dielectric constant induced by resonant radiation are detected as changes of the resonant frequency and/or merit factor.

For example in radio-frequency discharges the OG signal has been measured by way of different approaches: as a change in the current feeding the oscillator (as described in the paragraph II.a); by external coils inductively coupled to the plasma (Suzuki, 1981); or by internal electrodes (Labastie, 1982). A novel method for detecting atomic transitions in rf plasmas, very similar to the conventional OG method but not directly connected with the change in the discharge current, was introduced by Brandeberger (1987) (the optovoltaic effect). Recently we have developed a new scheme for impedance spectroscopy in which the physical parameter probed is not the current but the dielectric constant (Francesconi et al 1990). The sample is placed between the two parallel plates of a capacitor which is part of a resonant circuit at a frequency f ≈ 10MHz and with a given quality factor Q. A separate resonant circuit acts as a very stable reference (see Fig 12). Any change in the dielectric constant of the sample produces a change of both the merit factor, ΔQ and the resonant frequency, Δf. The reference resonant circuit gives a start signal while the laser perturbed circuit gives a stop signal. The temporal separation of these two signals is converted into a voltage signal and gives a measure of ΔQ and Δf. Higher sensitivity is achieved by using a phase-sensitive detection scheme. [A similar approach has also been developed by the group of Schawlow (Yan et al 1990). They operated a neon discharge bulb as a relaxation oscillator.]

We have tested our technique on several neon transitions by inserting part of a positive column in the "open" resonant circuit.

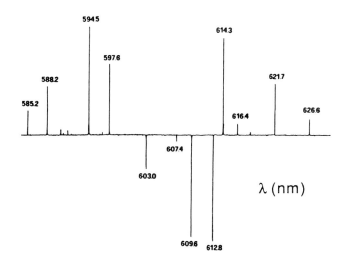

Figure 13   Broad band recording of the neon transitions in the R6G wavelength range using detection of the resonant frequency change, Δf.

Figure 13 shows a typical spectrum obtained by detecting Δf over the wavelength range of rhodamine 6G. Simultaneous recording of "conventional" OG signals exhibits the same relative intensities and signs for all the observed lines. Doppler-free spectroscopy of several neon transitions has also been successfully performed, obtaining a sensitivity one order of magnitude larger than the conventional OG spectroscopy. The physical mechanism leading to the formation of Δf and ΔQ signals are easily explained in terms of the dielectric constant (Lorentz formula):

$$\epsilon = [-m\omega_p 4p\ e^2/(\omega-ib)]\ \Delta n_e \qquad (3)$$

where ω_p is the plasma frequency and b is the electron collision frequency. Positive or negative signals in Fig 13 reflect the sign of the laser induced variation of n_e. This is the same result obtained in a positive column glow discharge where the discharge current is related to the electron density n_e by the relation $i = e \pi r^2 v_d n_e$, where v_d is the electron drift velocity and r the column radius. In our case, any changes in the real and imaginary parts of the complex dielectric constant correspond to Δf (dispersion) and ΔQ (absorption) respectively. To test this hypothesis we have applied this detection scheme to a sample of neutral sodium, exciting the D1 line by laser radiation. Strong Δf and ΔQ signals could also be observed in this case but only if the cell contained several Torr of neon buffer gas and the cell itself was heated to about 300 ºC. It is well known (Berman and Leventhal, 1978) that in this condition associative ionization is very effective in producing free charges.

Because of its wide applicability and non-intrusiveness, this technique could be particularly suitable, for example, in diagnosis of flames. A possible improvement could be achieved by increasing the resonant frequency because this corresponds to an increase in the quality factor Q.

5. CONCLUSIONS

In this paper the principal advantages and limitations of OG spectroscopy have been presented. These aspects have been discussed for specific cases concerning atomic species of interest in pure and applied physics as, for example, the oxygen atom. OG detection has resulted in very effective and easy monitoring of atomic oxygen in weak plasmas generated by rf excitation. Nevertheless, really high-resolution OG spectroscopy is seriously limited in comparison to other techniques here presented, such as polarization spectroscopy. OG spectroscopy has been improved by developing a new technique where the physical parameter probed is the dielectric constant of the medium under investigation (not necessarily a discharge).

ACKNOWLEDGMENTS

The author thanks all the colleagues that have made possible the realization of the experiments described in this paper. In particular he is grateful to M Inguscio for encouragement during the writing of the manuscript and G Tino for a critical reading of it.

REFERENCES

Barbieri B, Beverini N and Sasso A, Rev. Mod. Phys., in press.
Bearman G H and Leventhal JJ, Phys. Rev. Lett. 1978, 41, 1227.
Brandeberger J R 1987 Phys. Rev. A36, 76.
De Marinis E, Sasso A and Arimondo E 1988 J. Appl. Phys. 63, 649.
Demtroeder W 1988 Laser Spectroscopy, Springer Verlag, pp 505-514.
Doughty D K and Lawler J E 1983 Phys. Rev. A28, 773.
Ernst K and Inguscio M 1988 Rivista del Nuovo Cimento, Vol. II, N. 2.
Ernst K, Minutolo P, Sasso A, Tino G M and Inguscio M 1989 Opt. Lett. 14, 554.
Feld M S, Feldman B J and Javan A 1973 , Phys. Rev. A7, 257.
Francesconi M, Gianfrani L, Inguscio M, Minutolo P, Sasso A, Tino G M 1990
Appl. Phys. B51, 87.
Gianfrani L, Sasso A, Tino G M and Inguscio M 1990 Opt. Commun. 78, 158.
Goldsmith J E M, Ferguson A I, Lawler J E and Schawlow A L, 1979 Opt. Lett 4, 230.
Green R B, Travis J C and Keller R A 1976 Anal. Chem. 48, 1954.
Hänsch T W, Lyons D R, Schawlow A L, Siegel A, Wang Z Y and Yanh G Y 1981, Opt. Commun. 38, 47.
Hollberg L, Long-Shen Ma, Hohenstatt M and Hall J 1983 in Laser-Based Ultrasensitive Spectroscopy and Detection V, edited by Keller R A, Proc. Soc. Photo-Opt. Instrum. Eng. 426, 91.
Hurst G S, Payne M G, Kramer S D and Young 1979 Rev. Mod Phys. 51, 767.
Inguscio M, Minutolo P, Sasso A and Tino G M 1988, Phys. Rev. A37, 4056.
Kane D M 1984 J. Appl. Phys. 56, 1267.
Keller R A, Engleman R. Jr and Palmer B A 1980 Appl. Opt. 19, 836.
King W H 1984 Isotope Shift in Atomic Spectra, Plenum, New York.
Labastie P, Biraben F, Giacobino E, 1982 J. Phys. B15, 2595.
Lawler, J E, Ferguson A I, Goldsmith J E M, Jackson D J and Schawlow A L 1979 Phys. Rev. Lett. 42, 1046.
Maruyama Y, Suzuki Y, Arisawa T, Shiba K 1987 Appl. Phys. B44, 163.
Moore C E 1971 Natl. Stand. Ref. Data Ser. Natl. Bur. Stand. 35, Vols I, II and III.

Ochnin V N, Preobrazhenskii N G, Sobolev, N N and Ya Shaparev 1986 Usp. Fiz. Nauk. **148**, 473 (1986 Sov. Phys. Usp **29**, 260).
Penning F M 1928 Physica **8**, 137.
Smith P W and Hänsch W T 1976 Phys. Rev. Lett. **26**, 740.
Sasso A, Ciocca M, Arimondo E 1988a J. Opt. Soc. Am. **B5**, 1484.
Sasso A, Tino G M, Inguscio M, Beverini N and Francesconi M 1988b Il Nuovo Cim. **D10**, 941.
Sasso A, Minutolo P, Schisano M I, Tino G M, Inguscio M 1988c J. Opt. Soc. Am. **B5**, 2417.
Sasso A, Schisano M I, Tino G M, and Inguscio M 1990, J. Chem. Phys., in press.
Schenck P K, Travis J C, Turk G C 1983 J. Phys. **C7-44**, 23
Suzuki T 1981 Opt. Commun. **38**, 364.
Tino G M, Hollberg L, Sasso A, Inguscio M, Barsanti M 1990 Phys. Rev. Lett. **64**, 2999.
Wieman C and Hänsch T W 1976 Phys. Rev. Lett. **36**, 1170.
Yan G-Y, Fujii K I and Schawlow A L 1990 Opt. Lett. **15**, 142.

Frequency entrainment in the relaxation oscillator method of optogalvanic spectroscopy

J S Dunham, S C Bennett and C O Butler
Department of Physics, Middlebury College, Middlebury, VT 05753, USA

ABSTRACT: A simple neon bulb relaxation oscillator detection system for optogalvanic spectroscopy has been constructed. As an initial test of its performance in a magnetic field the detection system has been used to measure Zeeman effect absorption spectra for atomic Ne. The relaxation oscillator detector is a nonlinear dynamical system for which the phenomena of frequency entrainment and phase-locking can be demonstrated by illuminating the neon lamp with a chopped, resonant laser light beam. These phenomena appear to be well described by solutions to the van der Pol differential equation.

1. INTRODUCTION

Yan, Fujii and Schawlow (1990) recently reported a very sensitive technique for optogalvanic spectroscopy of atoms present in a low pressure gas discharge. In their method, a gas discharge lamp is operated as a relaxation oscillator and the effect of laser light tuned to an absorption line of atoms in the discharge gas is observed by measuring a change in the relaxation oscillator frequency. The relaxation oscillator frequency changes because the resonant light beam alters the excited state populations of the gas atoms, thereby changing the breakdown voltage of the discharge lamp. Using a simple neon bulb they recorded high resolution spectra of atomic Ne transitions with incident laser light intensities of less than 1 µW. The high sensitivity and experimental simplicity of the relaxation oscillator arrangement make this method attractive for the calibration and frequency stabilization of low power diode lasers.

The relaxation oscillator is a nonlinear dynamical system that was first studied in detail by van der Pol (1926,1927a,1927b) in a series of papers that have become classics in the field of nonlinear dynamics. Although van der Pol's earliest mathematical investigations concerned the passive electrical characteristics of a relaxation oscillator circuit, he extended his study to consider the active response of a relaxation oscillator to a small periodic electrical perturbation

© 1991 IOP Publishing Ltd

applied to the relaxation oscillator circuit. Van der Pol (1927b) showed that under suitable conditions a relaxation oscillator spontaneously changes from its natural oscillation frequency to the frequency of the periodic electrical perturbation in a process now known as frequency entrainment. This locking of the relaxation oscillator frequency to the frequency of an external periodic perturbation is a general phenomenon that is observed in many types of nonlinear dynamical systems.

In this paper we describe an optical extension of van der Pol's investigations. We report observations of frequency entrainment and phase-locking of a relaxation oscillator detector to a periodic *optical* perturbation produced by a chopped, resonant laser light beam incident on the neon bulb of the relaxation oscillator. The phenomena we observe are well described by solutions to the van der Pol differential equation.

2. EXPERIMENTAL APPARATUS

The neon bulb relaxation oscillator detection circuit used in the present work is shown in Figure 1 and is based on the description given by Yan *et al* (1990). In constructing the circuit we had success with a variety of commercial neon bulbs including the GE-2J neon lamp used by Yan *et al* (1990). The natural frequency of relaxation oscillations was set in the range 100-1000 Hz by selecting suitable values for the resistance R, capacitance C, and DC voltage V used in the relaxation oscillator circuit. The frequency of the relaxation oscillator was determined using a frequency-to-voltage converter that provided an analog signal to the A/D converter of a laboratory microcomputer where spectra were displayed and stored on disk for later analysis.

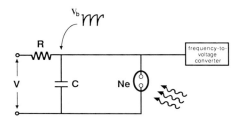

Fig. 1. Relaxation oscillator detection system for atomic Ne spectroscopy.

As a preliminary test of the relaxation oscillator detection system we investigated whether the neon lamp could be operated successfully in the presence of large magnetic fields such as those used for Zeeman effect measurements. This possibility is important because most optogalvanic cells are typically too large or too awkward in shape to be placed in the small homogeneous magnetic field region between the pole pieces of a large commercial electromagnet.

In a series of Zeeman effect measurements, a pair of permanent magnets of 19 mm diameter and 5 mm thickness were placed 15 mm apart to create a homogeneous magnetic field region of 0.31 T strength where the neon bulb was placed. We found that the relaxation oscillator frequency was altered only slightly by the presence of the 0.31 T magnetic field and that the stability of the oscillator was unaffected. Light from a Coherent 699-21 ring dye laser was passed through a polarizing system so that unpolarized, as well as linearly polarized, light beams of 50-100 mW intensity could be directed at the neon bulb.

In Figure 2 we give results of Zeeman effect measurements of the $2p_5 - 1s_5$, 588.2 nm transition in atomic Ne using our relaxation oscillator apparatus. For this transition the relaxation oscillator frequency decreases as the laser is scanned through resonance, as expected from the general arguments of Yan *et al* (1990). The linear polarization measurements show clearly the normal Zeeman effect splitting expected for this transition. We conclude that the relaxation oscillator method can be used to make Zeeman effect measurements at least at the 0.31 T magnetic field strengths investigated here.

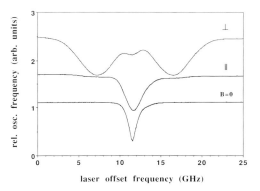

Fig. 2. Zeeman effect measurements of the 588.2 nm Ne transition for zero magnetic field (lower curve), linear polarization parallel to the magnetic field (middle curve), and linear polarization perpendicular to the magnetic field (upper curve).

3. FREQUENCY ENTRAINMENT

In van der Pol's original experiments on frequency entrainment (van der Pol 1927b), an external sinusoidal oscillator was placed in series with the neon bulb in a simple relaxation oscillator circuit such as that shown in Figure 1. When the external oscillator frequency was set near the natural frequency of the relaxation oscillator, he observed that even for very small amplitudes of the external oscillator signal, the frequency of the relaxation oscillator would spontaneously change to the frequency of the external oscillator. In addition to spontaneously changing its oscillation frequency, the relaxation oscillator became synchronous or phase-locked to the external oscillator. In the case that the external oscillator frequency was significantly different from the natural frequency of the relaxation oscillator, the relaxation oscillator would spontaneously change to a nearby harmonic or subharmonic of the external oscillator frequency. At still other frequencies the relaxation oscillator was unaffected or was found to operate in a completely erratic manner.

To analyze his experiments van der Pol modeled the neon lamp as a nonlinear circuit element with a simple functional form for the nonlinearity and the external oscillator was treated as a driving term to the usual relaxation oscillations. When put in dimensionless mathematical form, van der Pol's analysis leads to a study of the differential equation

$$\frac{d^2u}{dt^2} - \varepsilon(1-u^2)\frac{du}{dt} + u = B\cos\omega t \qquad (1)$$

where u is a dimensionless measure of the relaxation oscillator voltage, ε is determined by electrical parameters of the relaxation oscillator circuit, including the nonlinearity introduced by the neon bulb, and B and ω are the dimensionless amplitude and frequency of the external oscillator, respectively (Hayashi 1964). Equation (1) is known as the van der Pol differential equation and as a purely mathematical problem it has been treated in detail by Hayashi (1964). Numerical solutions of the van der Pol differential equation exhibit clearly the frequency entrainment and phase-locking phenomena that van der Pol observed in his original laboratory experiments.

In the present work we demonstrate experimentally that frequency entrainment and phase-locking can also be produced in a relaxation oscillator circuit by illuminating the neon bulb with a chopped, resonant light beam. It should be noted that this optical extension of van der Pol's original work is still largely electrical in character because the immediate effect of the chopped, resonant light is to alter the electrical characteristics of the neon bulb in a periodic way, resulting in behavior similar to the case when a periodic electrical oscillator is inserted in the circuit.

To demonstrate frequency entrainment and phase-locking we adjusted our relaxation oscillation detection system to a natural frequency of 400 Hz and then illuminated the neon bulb by a 100 mW laser beam resonant at the 588.2 nm Ne transition and chopped at frequencies in the range 100-2000 Hz. Figures 3(a)-3(d) show oscilloscope traces of the relaxation oscillator voltage and the chopped light signal for chopping frequencies of 328, 374, 404 and 452 Hz, respectively. It should be emphasized that the chopping signal serves as display trigger in all four oscilloscope displays, thus demonstrating frequency entrainment and phase-locking of the relaxation oscillator to the chopped light beam over the range of chopping frequencies from 325 to 460 Hz. No frequency entrainment phenomena were observed when the laser frequency was set well away from the 588.2 nm transition. We also investigated the effect of chopped, resonant light beams at chopping frequencies far from the

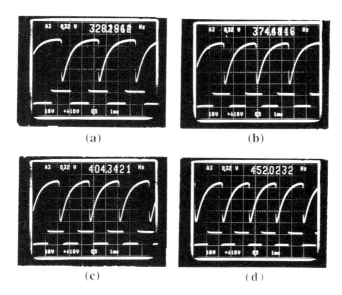

Fig. 3. Oscilloscope traces of the relaxation oscillator voltage (upper trace) and chopped light signal (lower trace/display trigger) for chopping frequencies of (a) 328, (b) 374, (c) 404, and (d) 452 Hz.

400 Hz natural frequency of the relaxation oscillator. Figures 4(a)-4(d) show oscilloscope traces of the relaxation oscillator voltage and the chopped light signal for chopping frequencies of 196, 803, 1192, and 1588 Hz. The 196 Hz case shown in Figure 4(a) corresponds to the case of relaxation oscillations locked to the first harmonic of the chopping frequency. The 803, 1192, and 1588 Hz cases correspond to relaxation oscillations locked to the first, second, and third subharmonics of the chopping frequency, respectively. Figures 4(a)-4(d) give clear evidence that frequency entrainment and phase-locking occur at both harmonics and subharmonics of the chopping frequency. The frequency range over which subharmonic and harmonic frequency entrainment could be established was typically 10 Hz, a range that is much smaller than the nearly 150 Hz over which entrainment to the fundamental was obtained.

In addition to the frequency entrainment and phase-locking phenomena described above we have also observed that for some chopping frequencies the relaxation oscillator appears to be chaotic; however, extensive measurements will be necessary in order to establish that truly chaotic behavior is indeed occurring.

4. CONCLUSIONS

We have demonstrated that the relaxation oscillator method can be used for Zeeman effect

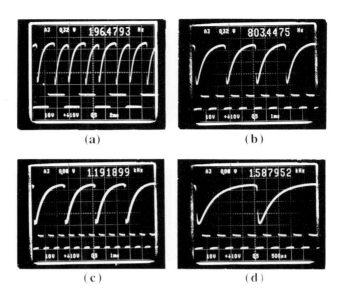

Fig. 4. Oscilloscope traces of the relaxation oscillator voltage (upper trace) and chopped light signal (lower trace/display trigger) for chopping frequencies of (a) 196, (b) 803, (c) 1192, and (d) 1588 Hz.

measurements and for demonstrating the nonlinear phenomena of frequency entrainment and phase-locking. We are currently engaged in numerical modeling of our relaxation oscillator detection system to gain a better understanding of the nonlinear phenomena described above. Although the expected nonlinear phenomena are clearly evident in the operation of our detection system, we are still working to find suitable approximations that can link various experimental parameters of our system to the specific mathematical form of the van der Pol differential equation. If we can establish such a link then we will be able to use a variety of mathematical treatments of van der Pol's differential equation to help identify new phenomena such as the possible chaotic behavior described above.

REFERENCES

Hayashi C 1964 *Nonlinear Oscillations in Physical Systems* (Princeton: Princeton Univ. Press) pp 285-308
van der Pol B 1926 *Phil. Mag.* **2** 763
van der Pol B 1927a *Phil. Mag.* **3** 65
van der Pol B and van der Mark J 1927b *Nature* **120** 363
Yan G-Y, Fujii K-I and Schawlow A L 1990 *Opt. Lett.* **15** 142

Isotope shift studies of Gd I transitions using laser optogalvanic spectroscopy

T Pramila

Department of Physics, IIT Kanpur, Kanpur 208016, India

ABSTRACT: Doppler limited laser optogalvanic spectroscopy measurements, using a cw ring dye laser pumped by an argon ion laser acting on a Gd/Ne hollow cathode discharge, have been carried out to determine the isotope shifts in 10 strong Gd I lines in the R6G spectral range. Due to the reduced Doppler widths of the spectral lines in hollow cathode discharges the isotope shifts for the more abundant even isotopes (160,158,156) contained in natural gadolinium could be obtained. The isotope shifts reported here are compared with the previously reported values obtained from conventional methods.

INTRODUCTION

The analysis of the spectra of the rare-earth atoms is made difficult by the strong interaction which exists among many configurations. It is very difficult to state correct designations and reliable wave functions, and hence there exist many discrepancies in the classifications of high lying levels of these elements given by different people (Martin *et al* 1978). In such cases, a systematic study of the hyperfine structure (hfs) and isotope shifts (IS) of the energy levels enables the unambiguous labelling of the energy levels. Moreover, from the analysis of the hfs and IS, detailed information can be extracted for electron charge densities at the nucleus in different configurations.

Gadolinium, which is a member of the rare-earth group of elements has several isotopic forms as given in Table 1 (Kropp *et al* 1985). A great

© 1991 IOP Publishing Ltd

Table 1 Abundance (%) of different isotopes in natural gadolinium

Isotope	160	158	157	156	155	154	152
Abundance	20.87	23.45	16.42	20.59	15.61	2.86	0.20

number of spectroscopists have been involved over a number of years in the study of the spectra of Gd I which resulted in 236 odd and 371 even energy levels being identified (Martin et al 1978). The study of the isotope shifts in Gd I spectra started with the work of Klinkenberg in 1946 (King 1943) and was extended by several workers (Brix & Engler 1952, Suwa 1953, Kopfermann et al 1957). In the recent past, Ahmad and coworkers (Ahmad et al 1976, 1979, 1982), using gadolinium samples enriched in ^{156}Gd and ^{160}Gd isotopes, have carried out an extensive study on the isotope shifts $\Delta T(156-160)$ of Gd I lines in the spectral range 393.0–497.5nm and reported the isotope shifts in the odd and the even energy levels involved in the transitions studied. Based on the observed shifts, the configuration assignments were confirmed or revised for some of the energy levels and in some cases the probable configurations for some unassigned levels were suggested.

In the present work, a study of the isotope shifts $\Delta T(160-158)$ and $\Delta T(160-156)$ for 10 Gd I lines in the spectral range 565–625nm was carried out using Doppler limited laser optogalvanic spectroscopy (LOGS). In LOGS the absorption of laser radiation by a sample under study, which is in the form of a discharge, is detected by measuring the change in the impedance of the discharge when it resonantly absorbs laser radiation. The gadolinium atoms for this study were obtained by sputtering from the cathode of a hollow cathode lamp. The sputtering process of sample preparation has the advantage of effectively populating the various energy levels on which spectroscopic studies can be carried out (Keller et al 1979). The high detection sensitivity of optogalvanic spectroscopy makes it convenient for the study of even the weakly populated states. Moreover the sputtered atoms, on collision with the buffer gas atoms, lose energy resulting in a low kinetic temperature and hence smaller Doppler width, thus making it possible to separate the contributions of different isotopes to the absorption profile, at least in some favourable cases. The isotope shifts obtained in the present work are compared with the values obtained from conventional spectroscopic techniques (Brix and Engler, 1952).

EXPERIMENTAL

The experimental set up used for the present measurements is shown in Fig. 1.

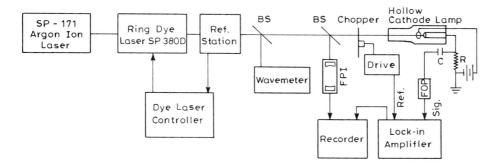

Figure 1   EXPERIMENTAL ARRANGEMENT

An externally stabilised single frequency ring dye laser (SP 380D, dye R6G) pumped by an Argon ion laser (SP 171), was the source of narrow line width (<10MHz) tunable radiation.   The peak output power was about 200mW. The dye laser can be electronically scanned over a 30GHz range.   The laser beam was chopped at a suitable frequency in the range 300-600Hz using an electro-mechanical chopper (Ithaco Model 220) for phase sensitive detection. A commercial Gd/Ne hollow cathode lamp (Instrumentation Laboratories Inc.) was used as a source of Gd atoms.   The Gd atoms in the discharge were obtained by cathode sputtering in the discharge and were at natural abundance. The lamp was operated at currents in the range 8-14mA using a stabilized d.c. power supply.   The dye laser beam was focussed into the cathode bore with the help of a lens.   The optogalvanic signal was picked up across a 40kΩ ballast resistor and was fed to a lock-in amplifier (Stanford Research Inc. Model SR510) through a 0.01 μF dc blocking capacitor and a fast over voltage protection circuit.   The signals were recorded on a strip chart recorder.   The high resolution scans were monitored by a 300 MHz free spectral range confocal Fabry-Perot interferometer (Coherent Inc. Model 216) which also provided the frequency calibration marks.   The dye laser was tuned to the desired gadolinium transition with the help of a wave meter (Burleigh Model WA-20).

RESULTS AND DISCUSSION

Figure 2 shows a part of the observed Gd/Ne laser optogalvanic spectrum

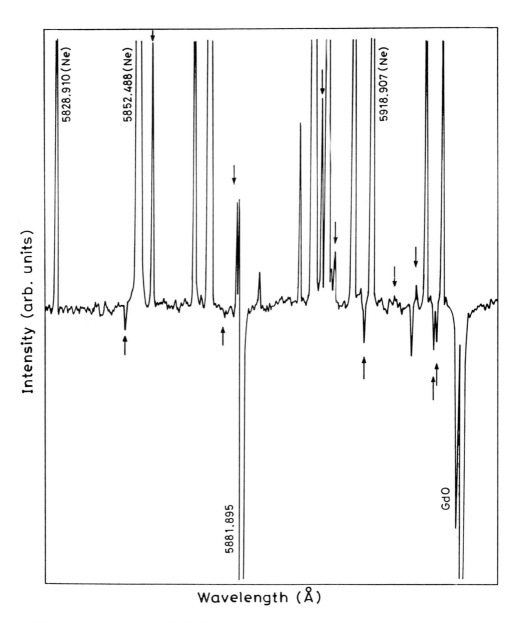

Figure 2   Broad Band LOGS Spectrum of Gd

which was recorded using a broad band dye laser (SP 375) operating in the 565–625nm region. This spectrum was calibrated with the help of the known neon lines. Gadolinium lines were identified based on the existing literature (Martin *et al* 1978, Russell 1950). Observed Gd I lines along with their relative optogalvanic intensities are listed in Table 2. As can be seen from Fig. 2, both types of transitions giving rise to increase as well as decrease in impedance are observed in gadolinium.

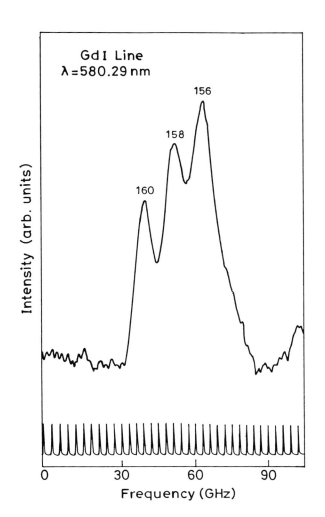

The isotope shifts for 10 of the observed transitions of Gd I in the region 565–625nm are studied. The wavelengths of the lines studied and the corresponding transitions are given in Table 3. The recorded LOGS profiles for the lines at 580.29nm and 574.6nm are given in Figs. 3 and 4 respectively. The line profiles are obviously dominated by the three even isotopes 156, 158 and 160 but the peak intensities do not reflect their natural abundance. This is because of the hyperfine structure of the two odd isotopes 155 and 157. The observed line profiles occur because the ^{157}Gd component which has negligibly small hyperfine structure,

FIGURE 3   LOGS PROFILE ON THE 580.29nm LINE

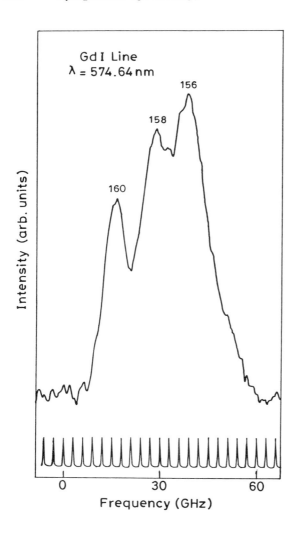

is superimposed on the ^{156}Gd component (Suwa1953) and hence the peak corresponding to ^{156}Gd gains in intensity. In some of the observed spectra a pronounced shoulder after the peak corresponding to ^{156}Gd is observed. This is due to the other odd isotope ^{155}Gd. Whenever the hyperfine structure splitting of this isotope is very small this shoulder appears and when the hyperfine structure is well spread the shoulder is absent (Suwa 1953). In the present study the contribution from the low abundance isotopes 154 and 152 could not be detected. In the present work, the IS values $\Delta T(160-158)$, and $\Delta T(160-156)$ only are reported. These values

FIGURE 4   LOGS PROFILE OF THE 574.64nm LINE

are obtained by treating the effect of the superposition of the hfs of ^{157}Gd on the position of the peak corresponding to the isotope 156 to be within the experimental errors (30 MHz) as these isotope shift values compare fairly well with the previously reported values (Table 4). All the lines studied in the present work involve transitions from the levels of the ground configuration (4f^{7}5d6s^2) to the levels of the configuration 4f^{7}5d6s6p. A study of isotope shifts of these levels using Doppler-free techniques like intermodulated optogalvanic saturation (IMOGS) and enriched samples of gadolinium would result in more accurate IS values for these levels which can be used to obtain

Table 2 LOGS spectra of Gd I along with the observed relative intensities and classifications.

Wavelength (nm)	Intensity (arb. units)	Transition (cm^{-1})
635.38	14.0	18070.26–33804.45
631.72	5.5	999.11–16824.60
631.64	2.5	7562.45–23389.80
629.29	13.0	999.11–16885.74
628.97	6.0	17909.94–33804.45
625.57	8.0	215.13–16196.00
623.47	2.0	6786.24–22820.91
622.44	13.0	0.00–16061.27
619.95	4.0	7103.46–23229.36
618.72	4.0	10883.56–27041.79
615.11	3.0	6976.49–23229.36
613.83	3.0	16775.02–33061.49
613.50	13.0	1719.06–18014.40
612.58	1.5	999.11–17318.94
611.41	sat	1719.06–18070.26
610.91	33.0	1719.06–18083.64
602.11	12.0	6786.24–23389.80
601.90	15.0	215.13–16824.60
599.91	16.5	6550.44–23215.07
599.67	7.5	18083.64–34754.96
598.45	13.0	215.13–16920.40
598.34	24.0	215.13–16923.38
597.73	29.0	6378.17–23103.64
596.32	13.0	7234.90–23999.92
593.77	17.0	6378.17–23215.07
593.68	20.0	6550.44–23389.80
593.03	10.0	6786.24–23644.19
592.33	7.0	8498.50–25376.34
591.68	17.0	7103.46–23999.92
590.84	23.0	0.00–16920.46
590.74	15.0	0.00–16923.38

Table 2 Continued

590.46	sat	999.11–17930.52
587.99	47.0	15174.00–32176.40
587.67	3.0	6378.17–23389.80
587.06	5.0	999.11–18070.26
585.62	sat	999.11–18083.64
584.85	11.0	6550.44–23644.19
584.09	6.0	10359.94–24475.74
582.40	27.0	215.13–17380.83
580.92	5.0	15852.25–33061.49
580.77	6.0	6786.24–23999.92
580.29	25.0	0.00–17227.96
579.14	53.0	532.98–17750.00
575.32	10.0	10359.94–27736.68

sat – saturated

Table 3 Lines of Gd I for which an IS is obtained

Serial No	Wavelength (nm)	Energy levels (cm^{-1})	
		Odd	Even
1	569.62	532.98(a $^9D_4$)	18083.64(z $^9D_5$)
2	574.64	532.98(a $^9D_4$)	17930.52(z $^9D_4$)
3	575.19	0.00(a $^9D_2$)	17380.81($^9F_2$)
4	579.14	532.98(a $^9D_4$)	17795.27(z $^9D_3$)
5	580.29	0.00(a $^9D_2$)	17227.97($^9F_1$)
6	582.40	215.12(a $^9D_3$)	17380.81($^9F_2$)
7	585.62	999.12(a $^9D_5$)	18070.26(z $^9D_6$)
8	590.46	999.12(a $^9D_5$)	17930.52(z $^9D_4$)
9	601.90	215.12(a $^9D_3$)	16824.60(z $^{11}D_4$)
10	611.41	1719.09(a $^9D_6$)	18070.26(z $^9D_6$)

Table 4 Experimental Isotope Shifts of Gd I lines Studied

Wavelength	Isotope shifts (MHz)			
	ΔT(156−160)		ΔT(160−158)	
	Present	Ref 4	Present	Ref 4
569.62	2250	2160	1200	1050
574.64	2250	2160	1200	1080
575.19	2160	2160	1050	1080
579.14	2250	2190	1350	1170
580.29	2400	2250	1280	1080
582.40	2100	2220	1200	1110
585.62	2100	2160	900	1080
590.46	2200	2280	1200	1140
601.90	2250	2250	1050	1110
611.41	2100	2160	900	1050

the term and J-dependent effects in the optical isotope shifts by applying the PADIS method given by Bausche (Bausche 1969).

To conclude, the LOGS technique has been shown to be very simple and convenient for high resolution spectroscopic studies of rare earth elements like gadolinium which are characterised by very dense and complex spectra.

Acknowledgments

I am grateful to the Department of Science and Technology, New Delhi, for the financial aid under the Project No: SP/YS/PO8/86 to carry out this work.

References

Ahmad, S.A., Saksena, G.D. and Venugopalan, A., Physica 81C, 366 (1976).
Ahmad, S.A., Venugopalan, A. and Sakena, G.D., Spectrochim. Acta 34B, 221 (1979)..
Ahmad, S.A., Venugopalan, A. and Sakena, G.D., Spectrochim. Acta 37B, 637 (1982).
Bauche, J., Physica, 44, 291, (1969).

References (Contd)

Brix, P. and Engler, H.D., Z. Phys. 133, 362, (1952).
Keller, R.A., Engleman, R. Jr. and Zalewski, E.F., J. Opt. Soc. Am. 69, 738 (1979).
King, A.S., Astrophys. J., 97, 323 (1943).
Kopfermann, H, Kruger, L. and Steudel, A., Ann. Phys. 20, 258, (1957).
Kropp, J.-R., Kronfeldt, H.-D. and Winkler, R., Z. Phys. A-Atoms and Nuclei 321, 56 (1985).
Martin, W.C., Zalubas, R and Hagen, L. (Eds) Atomic Energy Levels, The Rare-E Elements, Natl. Stand. Ref. Data Ser., Natl. Bur. Stand (U.S.) 60, (U.S. G.P.O., Washington D.C., 1978).
Russell, J. Opt. Soc. Am., 40, 550 (1950).
Suwa, S., J. Phys. Soc. Jpn. 8, 377 (1953).

Optogalvanic spectroscopy in commercially available hollow cathode lamps

M.Duncan and R.Devonshire, High Temperature Science Laboratories, Chemistry Department, Sheffield University, Sheffield S3 7HF, U.K.

ABSTRACT. This paper contains two parts, A and B. Part A describes an optogalvanically based wavelength calibration system developed for a newly installed pulsed dye-laser. A commercially available "see-through", mixed gas, hollow cathode lamp, a DC stabilised bench power supply and a solid etalon form the basis of the calibration system. Part B describes experiments which exploit the calibration system to identify resonant two-photon transitions originating from the 6^3P_J excited states of mercury. The transitions are detected optogalvanically within a hollow cathode lamp of similar design to that used for wavelength calibration.

Two applications of optogalvanic (OG) spectroscopy in commercially available hollow cathode discharge lamps are presented in parts A and B of this paper. Part A describes an optogalvanically based wavelength calibration system developed for the newly installed pulsed dye-lasers. Part B describes an investigation into the multiphoton excitation of the 6^3P_J excited states of mercury.

PART A. OPTOGALVANIC WAVELENGTH CALIBRATION SYSTEM

A.1 INTRODUCTION

The recently installed pulsed dye-lasers required a simple wavelength calibration system accurate to better than 0.05 cm^{-1} which would be useful over the broad wavelength range available with the lasers. A number of possible methods were considered including the purchase of a pulsed wavemeter, the use of an iodine cell for the saturated laser absorption (or for fluorescence emission) and the use of OG spectroscopy. The chosen calibration device was based upon the OG effect because of the overwhelming advantage of the technique over conventional calibration methods in the ease with which calibration spectra may be obtained.

Wavelength calibration is essential when using dye lasers because of the mechanical inadequacies, and the dye solution instabilities, which result in random and systematic discrepancies between the actual wavelength of the dye laser output and the indicated wavelength. The calibration of tunable dye lasers using the OG effect is a well established technique. King et al (1977, 1978) described the use of OG spectroscopy in a Ne/Na hollow cathode lamp as a method of direct calibration of laser wavelength using a chopped, tunable, CW laser. Further calibration spectra have been recorded using lamps of various types and fill. Rosenfeld et al (1979), Nestor (1982), Juguang et al (1982), Hamer and Spoonhower (1984), Su et al (1987) have all used lamps containing either neon or argon. Neon and argon gaseous discharges have been employed most extensively for calibration purposes because these gases are commonly used as buffer gases within hollow cathode lamps and they provide an acceptable density of calibration lines.

A.2 EXPERIMENTAL

A schematic of the recently installed pulsed laser system is shown in **Figure 1**. The two dye lasers (Spectra-Physics PDL3's) are pumped by either the second or third harmonic output of the Nd:YAG laser (Spectra-Physics single-mode, Gaussian optic DCR-3), to provide independently tunable radiation from 380 to 960 nm. This wavelength range may be further extended using a Raman shifter, frequency doubling, etc. The linewidth of the dye lasers is specified to be 0.05 cm^{-1}, and to date has been measured to be <0.1 cm^{-1}. The temporal FWHM of the 20 Hz laser pulses is approximately 7 ns.

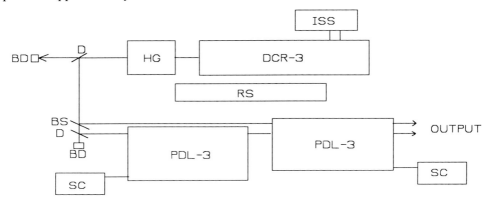

Figure 1. The pulsed laser system.

BD Beam Dump, **BS** Beam Splitter, **D** Harmonic Separating Dichroic, **DCR-3** Nd:YAG laser (Spectra-Physics single-mode Gaussian optic), **HG** Harmonic Generating Crystals, **ISS** Injection Seeding System, **PDL-3** Pulsed Dye-Laser (Spectra-Physics), **RS** Raman Shifter, **SC** Questek Scan Controller

The probe beam for calibration was prepared by sampling (achieved using a simple glass plate) and aperturing a small percentage of the scanning dye-laser beam, to generate a collimated beam with a diameter of about 1 mm and a pulse energy of typically 0.05 mJ. After appropriate attenuation the beam was then passed through the horizontally positioned cathode of a Hamamatsu mixed gas fill hollow cathode lamp containing a specified 2 Torr of argon and 3 Torr of neon (see **Figure 2**). The aluminium cathode was "see-through" providing an optically clear path for the dye-laser beam. The lamp was mounted upon a protective stand compatible with other optical equipment in the laboratory and constructed in our workshops. It was run at 17 mA using a Farnell DC stabilised power supply (Type E.350 Mk-II) and an appropriate ballast. The OG spectra were recorded using an RC differentiating circuit (with a time constant of about 10^{-6} s) and a boxcar (SRS Model 250) operating in single point averaging mode. The boxcar was triggered using either one of the synchronous outputs from the Nd:YAG laser or a photodiode. Data acquisition and treatment was controlled with an IBM compatible PC running SRS software.

A single, uncoated, parallel-faced silica disc was used as a low-finesse (approximately 2.5) reflection etalon (Seltzer et al 1988). When the laser beam is incident on the solid etalon at a small angle return reflections occur from the front and back faces and subsequently interfere. If the interference pattern is viewed through a pin hole (diameter 10^{-4} m) by a photodiode, and the dye-laser is scanned, a strong intensity modulation is detected. The beam was prepared by sampling the dye-laser output using the reflection from a single face of a quartz prism, thereby avoiding additional, confusing interference effects encountered when a glass plate sampler is used. The etalon interference fringes were recorded simultaneously with the OG calibration spectra using an additional boxcar module. When calibrated, the regularly spaced fringes can be used to interpolate between a reference spectral feature and an unknown wavelength of interest, leading, thereby, to a precise determination of the unknown wavelength.

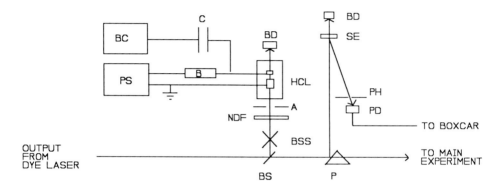

Figure 2. Schematic of the apparatus used for the wavelength calibration system.

A Aperture, **B** Ballast Resistor, **BC** Boxcar, **BD** Beam Dump, **BS** Beam Sampler, **BSS** Beam Steering System, **C** Capacitor, **HCL** Hollow cathode lamp, **NDF** Neutral Density Filter, **P** Prism, **PD** Photodiode, **PH** Pin Hole, **PS** Power Supply, **SE** Solid Etalon

A.3. RESULTS AND DISCUSSION

OG calibration spectra have been recorded using a variety of laser dyes over a wavelength range covering approximately 13,500 - 24,000 cm^{-1} (740 - 420 nm). In these spectra virtually all expected Ar I transitions have been observed with linewidths of typically 0.3 cm^{-1}. Useful wavelength tables for argon (and other elements) are given by Phelps (1982), Minnhagen (1973) and Striganov and Sventitskii (1968). For cross reference, tables of term values have been prepared by Moore (1949) and updated by Bashkin and Stoner (1975). Signals originating from neon transitions have been observed in some wavelength regions and are still under investigation by us. No Ar II or Al I transitions have been observed to date.

The etalon reflection interference fringes are approximately sinusoidal in appearance. To determine the etalon fringe separation two Ar I spectral lines of known wavelength may be used. The line centres of the two chosen spectral features is precisely determined by interpolating between individual boxcar data points. The calculated wavelength difference between the two lines is then divided by the exact number (including fraction) of etalon measured fringes to determine the fringe separation in cm^{-1} in that region of the spectrum. **Figure 3** shows an OG calibration spectrum and etalon trace in the wavelength range 16,010 - 16,046 cm^{-1}. The Ar I transitions observed in this wavelength range are the

4p[1/2]1 -od[1/2]0 and 4p[3/2]2-5d[5/2]2 (Paschen notation). The wavelengths of these transitions in vacuum are quoted by Minnhagen (1973) as 16,012.510 cm^{-1} and 16,044.552 cm^{-1}. Using these values the fringe separation was found to be 0.793 cm^{-1}. Using this fringe separation and a single nearby Ar I transition of accurately known wavelength, it is then possible to determine the wavelength of an unknown spectral feature with a certainty of at least $\pm$ 0.05 cm^{-1}, which corresponds to the dye-laser output bandwidth.

At higher laser powers (typically 0.5 mJ) the observed Ar I line profiles exhibit distortions which were initially believed to be saturation effects. However, upon focusing the beam and thereby limiting the interaction volume between the beam and discharge plasma to a narrow track along the axis of the lamp these distortions were found to disappear. As a result these observations are believed to relate to spatial effects within the discharge; it is well known that excited states play a particularly important role at the surface of the cathode in hollow cathode lamps (Uchitomi 1983). When the laser beam is focused, additional OG signals are observed which result from multiphoton transitions. Such multiphoton transitions in inert gases have been previously observed by Bickel and Innes (1985). Other signals have also been observed which have not been identified as yet. They do not correspond to known argon, neon or aluminium (cathode material) transitions and are still under investigation.

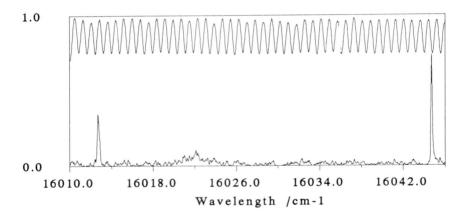

Figure 3. OG calibration spectrum and etalon trace in the wavelength range 16,010-16,046 cm^{-1}.

PART B. OPTOGALVANIC DETECTION OF RESONANT MULTIPHOTON TRANSITIONS FROM THE 6^3P_J STATES OF Hg I

B.1 INTRODUCTION

The low pressure mercury-argon discharge has been well characterised compared to many other electrical discharge systems. Progress in low pressure mercury-rare-gas discharge research has recently been reviewed by Maya and Lagushenko (1990). The particular knowledge of this discharge type is mainly because of its great interest to the lighting industry.

A number of laser spectroscopic methods have been used to study this discharge. For example van der Weijer and Cremers (1986, and references therein) have used laser induced fluorescence (LIF), optogalvanic spectroscopy (OGS) and laser induced stimulated emission (LISE) as low pressure mercury discharge diagnostics. OGS was used also in the studies of Reddy et al (1989) and of Richardson et al (1987). In the work of Richardson and his collaborators this discharge type was the basis for an investigation into the OG effect itself.

The 6^3P_J mercury states play a particularly important role within the mercury discharge. The 6^3P_1 state is the origin of the 254 nm resonance line, which is converted into visible radiation by the phosphor coating on the inner wall surface of fluorescent lamps. The population of the 6^3P_1 state is closely coupled to the 6^3P_2 and 6^3P_0 metastable states through electron collisions. Hence the radial densities of these levels are crucial in determining the radiation transport of the 254 nm line to the edge of the tube. Consequently, spatially resolved measurements of the excited-state densities are important for the verification of predictions by, and/or improvements to, the many computational models of this discharge (see for example the investigations of Bigio (1988) and Moskowitz (1987)).

Even when only a single probe beam is involved, multiphoton spectroscopy offers, by its very nature, spatially resolved measurements of number densities. This is because, in order to generate the large flux densities required to stimulate the multiphoton transitions the beam must generally be focused. As a result the multiphoton induced signals only arise in the focal region of the beam. The region of signal generation can be realistically simulated using the model of Kaabar and Devonshire (1991). To date, however, there have been no reports of the use of multiphoton spectroscopy as an active discharge diagnostic in operating light sources. We are presently identifying multiphoton laser induced fluorescence schemes involving Hg excited states which will form the basis for spatially resolved measurements of their relative density distributions (and that of related species).

The sensitivity and ease of detection offered by OG spectroscopy provides a suitable starting point for the development of such multiphoton diagnostic techniques for both low and high pressure mercury discharges.

B.2 EXPERIMENTAL

We have detected resonant two-photon transitions in a mercury-dosed Hamamatsu "see-through" hollow cathode lamp of similar design to that used in part A. This lamp was run at a current of less than 1 mA using a stabilised Brandenburg DC power supply (model 75R; photomultiplier tube power supply) and a suitable ballast. The lamp contains an argon and neon mixed gas buffer. The laser induced OG signals were detected and recorded using circuitry and detection apparatus identical to that described in Part A.2. of this paper. All three output signals, mercury OG spectrum, OG calibration and reflection interference fringes were detected and recorded simultaneously.

The dye-laser beam, after sampling for calibration, contained typically 90% of the dye laser output. This beam was attenuated (dependent upon the particular dye efficiency and in the proportion of the Nd:YAG harmonic used to pump the dye-laser) to provide a linearly polarised beam with a diameter of approximately 7 mm and a pulse energy of 0.3 - 0.4 mJ. This was focused with a 150 mm focal length, best form, lens, providing an estimated beam waist diameter of 15 microns (three times the diffraction limited beam waist of 5 microns) over the wavelength range used and a peak power density of approximately 300 MWcm^{-2}. The hollow cathode lamp was positioned so that the beam waist was located axially near the centre of the 19 mm long hollow cathode.

Two-photon allowed transitions were stimulated from each of the 6^3P_J states to their corresponding 7^3P_J states, i.e. J" to J' = 0 (see **Figure 4**). To generate the dye-laser beams at the required wavelengths two dyes in solutions of methanol were used, LDS 750 (range 14,000 to 12,600 cm^{-1}) and DCM (range 16,500 to 14,900 cm^{-1}).

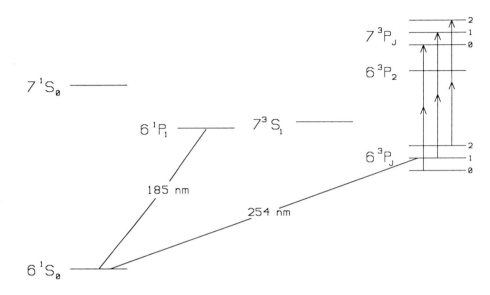

Figure 4. Simplified energy level diagram for mercury.

B.3 RESULTS AND DISCUSSION

The wavelengths of the two-photon allowed transitions can be calculated from the term values given by Moore (1949). These term values are quoted to within ± 0.01 cm^{-1}. The calculated resonant two-photon frequencies for the 6^3P_J to 7^3P_J (J"- J' = 0) transitions are 15,935.79 cm^{-1} (J"=J'=0), 15,124.80 cm^{-1} (J"=J'=1) and 13,582.27 cm^{-1} (J"=J'=2). All three of these two-photon transitions have been observed in the hollow cathode lamp. **Figure 5** shows the 6^3P_0 to 7^3P_0 transition, together with the corresponding region of the Ar I calibration spectrum, and the etalon trace. The observed Ar I line originates from the 4p[1/2]2 to 5d[7/2]3 transition (15,922.5984 cm^{-1}). Using this, and other, Ar I lines the position of the 6^3P_0 to 7^3P_0 transition was measured to be 15,935.81 cm^{-1} i.e. within 0.02 cm^{-1} of the calculated value, which is as good agreement as could be hoped for, given that the dye-laser bandwidth is 0.05 cm^{-1}. The multiphoton mercury signals are not observed when the laser is unfocused but the energy per pulse is maintained. Transitions arising from the argon neon gas fill are weakly observed within the mercury dosed hollow cathode lamp.

The two-photon signals originating from the 6^3P_0 and 6^3P_2 metastable states are strong and negative (i.e. decrease in the voltage across the discharge). This is consistent with a resonant two-photon absorption, followed by a single-photon ionisation from the upper state. Goldsmith (1983) observed similarly large signals using resonant multiphoton optogalvanic detection of atomic oxygen in flames. The large OG signal strength resulting from these transitions may provide the basis for new laser diagnostic techniques for operating Hg discharge lamps.

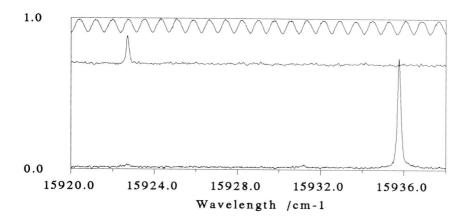

Figure 5. Two-photon OG spectrum of mercury in the wavelength range 15,920 to 15,938 cm^{-1} (lower trace), Ar I calibration spectrum and etalon trace is also shown.

SUMMARY

This paper has described two applications of OG spectroscopy in commercially available hollow cathode discharge lamps.

A mixed gas fill hollow cathode lamp has been used as the basis of a wavelength calibration system. It is providing us with a simple and versatile method for the determination of unknown wavelengths over a broad wavelength range with an accuracy that is well matched to the dye-lasers' characteristics.

A mercury dosed commercial hollow cathode lamp has also been used to identify resonant multiphoton transitions from the 6^3P_J mercury excited states. The work is viewed as the first step in the development of cross beam multiphoton diagnostics for low and high pressure mercury discharge lamps. A complete account of this work is to be published (Duncan and Devonshire 1991).

Further investigations for the application of multiphoton spectroscopy as a diagnostic within light sources are under way within our laboratories.

ACKNOWLEDGEMENTS

The authors would to thank D.O. Wharmby for useful discussions throughout the course of this work.

MD acknowledges financial support from the S.E.R.C. and THORN Lighting Ltd., (Light Sources Division).

REFERENCES

Bashkin S and Stoner J O 1975 Atomic Energy Levels and Grotrian Diagrams
    (North-Holland: Oxford)
Bickel G A and Innes K K 1985 Appl. Opt. **24(21)** 3620
Bigio L 1988 J. Appl. Phys. **63(11)** 5259
Duncan and Devonshire (to be published)
Juguang J, Changtai J and Songyue W 1982 J. Laser Phys. (China) **9(10)** 657
Goldsmith J E M 1983 J. Chem Phys. **78(3)** 1610

Hammer C A and Spoonhower J P 1984 Appl. Spec. **38(2)** 212
King D S, Schenck P K, Smyth K C and Travis J C 1977 Appl. Opt. **16(10)** 2617
Kaabar W and Devonshire R 1991 (to be published)
King D S and Schenck P K 1978 Laser Focus
Maya J and Lagushenko R 1990 Advances in Atomic, Molecular and Optical Physics **26** 321
Minnhagen L 1973 J. Opt. Soc. Am. **63(10)** 1185
Moore C E 1949 Atomic Energy Levels, (NBS: US)
Moskowitz P E 1987 Appl. Phys. Lett. **50(14)** 891
Nestor J R 1982 Appl. Opt. **21(22)** 4154
Phelps F M 1982 MIT Wavelength Tables Vol. 2: Wavelengths by Element, (MIT Press: England)
Reddy B R Venkateswarlu P and George M C 1989 Opt. Comm. **73(2)** 117
Richarson W H, Maleki L and Garnire E 1987 Phys. Rev. A **36(12)** 5713
Rosenfeld A, Mory S and Konig R 1979 Opt. Comm. **30(3)** 394
Seltzer M D, Piepmeier E H and Green R B 1988 Appl. Spec. **42(6)** 1039
Striganov A R and Sventitshkii N S 1968 Tables of Spectral Lines and Neutral and Ionised Atons, (New York: Plenum)
Su M C, Oritz S R and Monts D L 1987 Opt. Comm. **61(4)** 257
Uchitomi N, Nakajima T, Maeda S and Hirose C, 1983 Opt. Comm. **44(3)** 154
van de Weijer P and Cremers R M M 1985 Radiative Processes in Discharge Plasmas, Ed by J.M. Proud and L.H. Luessen, NATO ASI Series B:Physics Vol. 149, pp 65-93

Concentration profiling in a neon optogalvanic lamp using concentration-modulated absorption spectroscopy

T R Griffiths, W J Jones and G Smith

Department of Chemistry, University College of Swansea, Singleton Park, Swansea, West Glamorgan, SA2 8PP.

ABSTRACT: The quantitative pump-probe spectroscopic technique Concentration-Modulated Absorption Spectroscopy, in which a modulated pump laser perturbs the population difference between two electronic states interrogated by a probe laser of the same wavelength, is employed to determine absolute species concentrations and state lifetimes for the four lowest excited states of neon in a hollow-cathode discharge lamp. The variations of the species concentrations in localized regions of space are determined and information is derived on the kinetic excitation and de-excitation rate processes that determine the species concentrations in the various levels.

1. INTRODUCTION

Laser spectroscopic methods are especially appropriate for the analysis of harsh chemical environments where other more conventional techniques are unsuitable, the only requirements being transparent input and output ports for the laser radiation, a high degree of transmission through the sample and a suitable optical transition. For the most part these spectroscopic methods take advantage of direct optical absorption or emission processes between discrete pairs of energy states or inelastic scattering processes such as Raman spectroscopy in its various linear and non-linear guises. However, the most powerful generally-applicable of these methods, absorption spectroscopy, even with laser sources is of limited value with direct detection of the species absorbance as a result of low sensitivity, poor spatial resolution, and the inadequacy of concentration and temporal measurements, so that major efforts have been devoted to developing other detection methods to overcome some of these shortcomings. For the most part these other methods have taken advantage of the high photon flux available with monochromatic laser sources to perturb the population differences between pairs of coupled states, the laser-induced perturbations being detected most sensitively by indirect detection methods such as opto-galvanic or opto-acoustic spectroscopy or by laser-induced fluorescence.

A detection method which also relies on the perturbation of the population difference between states coupled by the radiation field is photo-induced bleaching or saturation

© 1991 IOP Publishing Ltd

spectroscopy, although this method is also of low sensitivity for the measurement of the primary absorption process when used with low-repetition-rate pulsed lasers. An approach to the sensitivity disadvantage of pump-probe laser absorption spectroscopy is to impress an amplitude (or polarization) modulation on the pump laser and to abstract the small amount of modulation transmitted to the probe through the optical mixing process by means of an appropriate lock-in-amplifier. Used with a very rapid modulation frequency (~ 10 MHz) this method is capable of recording fractional intensity changes of the probe laser of as small as 10^{-8} (shot-noise-limited performance for ~ 6 mW of probe laser power at 600 nm incident on a photodiode detector of 50% quantum efficiency), although this exceptional sensitivity is degraded at the lower modulation frequencies that sometimes prove to be necessary for the study of species with long-lived (> 100 ns) excited states. This sensitivity is not quite as great as is attainable with background-free techniques, such as opto-galvanic spectroscopy, but there are some compensating advantages which, when considered in conjunction with the exceptional sensitivity, make this a very powerful method for the study of atomic and molecular species in all states of aggregation, including excited state species such as occur in electrical discharges or in flames.

The characteristics of this approach to the study of absorption spectroscopy, termed Concentration-Modulated Absorption Spectroscopy, have been placed on a formal quantitative footing in a series of publications that consider the use of this method with both continuous-wave and mode-locked picosecond pulsed laser systems (Langley et al. 1985; Langley et al. 1986; Beaman et al. 1986; Jones 1987; Mallawaarachchi et al. 1987; Czarnik-Matusewicz et al. 1988).

When coincident with the frequency of a transition between two discrete energy states a pulse of pump laser radiation containing i_p photons experiences a net absorption as species are excited from the more populated to the less populated state. If this absorption occurs in a highly localized region of space, such as at the focus of the laser beam, the resulting population perturbation can lead to a large fractional change in the concentration difference between the two levels and thence to a significant diminution of the absorption of a succeeding probe laser pulse tuned to this same transition. When both probe and pump pulses are focussed together into a sample, avoiding significant saturation, integration of the incremental changes of probe laser transmission (Langley et al. 1986) over the complete focal zone yields an expression for the "gain", G, (attenuated absorption) of the probe laser which is given by:-

$$G = \frac{2\pi . \sigma^2 . N\Delta . i_p}{\lambda}. \tag{1}$$

$\Delta (=(N_1 - N_2)/N)$ is the equilibrium fractional population difference between levels 1 and 2, N the total concentration (atom cm^{-3}), σ is the absorption cross-section and λ the pump/probe laser wavelength. This expression is extremely interesting because the cross-section, σ, and $N\Delta$ ($\simeq N_1$) exhibit a different power dependence, whereas in the Beer-Lambert expression,

$$\frac{1}{\ell} \cdot \ln(I_o/I) = \sigma \cdot N\Delta \tag{2}$$

they display the same power dependence. As a result, if both the absorbance and the "gain" can be measured for a sample, known or unknown, of length ℓ it is possible to combine equations (1) and (2) to determine the species concentration

$$N\Delta = \frac{2\pi \cdot i_p \cdot [\ln(I_o/I)]^2}{\ell^2 \cdot \lambda G} \tag{3}$$

and the absorption cross-section, σ.

There are, of course, limitations on the use of the technique. For example, saturation effects need careful consideration (Jones 1987). These limitations are no more restrictive than for many other scientific methods, and given the ability to measure the "gain" and the absorbance for the same sample it is a very powerful method of measuring species concentrations even for samples present at high dilution. Further, this information can be provided for species present in highly localized regions of space (comparable with the dimensions of the confocal zone of a laser beam) so that it becomes possible to determine absolute concentration profiles for inhomogeneous samples.

An additional feature of the COMAS method is the ability to determine the lifetimes of electronic states by delaying the incidence of the probe pulse by known or varying time delays following excitation by the pump. After its initial creation, the population difference between levels 1 and 2 decays to its equilibrium value at a rate which is determined by the decay and repopulation kinetics of these levels, so that the probe transmission rises toward its transmission in the absence of the preceding pump pulse and the "gain" tends to zero. By monitoring the "gain" as a function of the probe time delay it is thus possible to determine the lifetimes in levels 1 and 2 (Beaman et al. 1986) an extension of this approach for longer-lived species being possible (Mallawaarachchi et al. 1987; Czarnik-Matusewicz et al. 1988) by varying the modulation frequency, and state lifetimes in the range $10^{-11} - 10^{-2}$ s are measurable by one or other of these two approaches.

The theoretically-derived "gain" expressions were confirmed in the above studies by investigations of the spectra of a variety of atomic and molecular species in flames and in the solution and solid states and the extent of the applicability of the method was examined in full. Following completion of this essential introductory work these methods have been applied to a variety of chemical systems to which they are ideally suited. One of these, which is of particular interest to this meeting relates to the investigation of the four 1s states of neon excited in an opto-galvanic discharge lamp and it is this study which will be reported in some detail in this paper.

2. THE EXPERIMENTAL STUDY

Excitation of the four 1s states of neon was examined in a Hamamatsu Laser Galvatron Lamp, Model L2783-25. This see-through hollow cathode lamp was operated from a ripple-free, stable source of 260V, the d.c. current through the lamp being limited to the range 0.03 - 10 mA by a variety of resistors, across which the opto-galvanic signal was generated. The hollow cathode lamp itself is a hollow tube of ~ 2.7 mm diameter and of length ~ 17.5 mm, the cathode being displaced by ~ 5 mm from the concentric ~ 3 mm long ring anode. With the electrodes located axially between a pair of optical windows it was relatively straightforward to focus the laser beams together into the lamp and to investigate the spatial distribution of species by moving the lamp radially and longitudinally in relation to the beam focus.

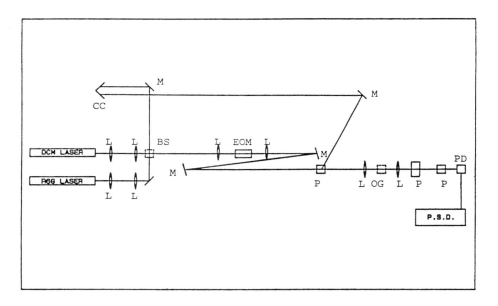

Figure 1. Plan layout for the experimental studies in Concentration-Modulated Absorption Spectroscopy.
L - Lens; CC - Corner Cube Retroreflector; EOM - Electro-optic Modulator; P - Polariser; OG - Optogalvanic Discharge Lamp; PD - Photodiode; BS - Beam splitter.

The experimental layout of the mode-locked lasers and the optical system is displayed in Figure 1, the lasers being arranged so that either the DCM or R6G laser may be employed as the pump laser in association with either laser as the probe: in the COMAS method one laser acts as both pump and probe. Both lasers consist of a continuous train of pulses of duration ~ 10 ps separated by 12.2 ns. Lifetimes were determined by monitoring the probe "gain" as the corner cube retro-reflector was scanned in time from pre-pulse-coincidence to a probe pulse delay of ~ 1 ns. Because this study dealt entirely with transitions among the energy states of atoms, pump and probe beams could be combined in orthogonal polarization at a polarizer, the pump beam being rejected satisfactorily post-sample by an analyzing polarizer before transmitting the probe beam to the photodiode detector for subsequent analysis. The measured "gain" is obtained from the rectified signal voltage from the lock-in amplifier after correcting for the lock-in-amplifier gain and the d.c. signal voltage obtained directly from the photodiode.

3. EXPERIMENTAL RESULTS

With either the DCM or the Rhodamine 6G laser incident on the opto-galvanic lamp it proved possible to record the optogalvanic spectra readily by chopping the laser beam before the discharge lamp and transmitting the alternating voltage developed across the series resistor to a suitable lock-in-amplifier. Though these spectra were satisfactory, they in no way reflected the sensitivity of the opto-galvanic technique because the line width of the picosecond lasers (1 - 2 cm^{-1}) was some 20 times the line width of the neon transitions. This factor had a significant bearing on the method of measuring the species concentrations because of the difficulty of measuring the direct absorbance with such broad line-width lasers. The concentration-modulated (COMAS) spectra were measured readily even with the picosecond lasers because of the sensitivity of the method, as shown in Figure 2(B) which displays the COMAS signals obtained with a 10 mA discharge current as the picosecond DCM laser was scanned over part of its operating range.

This spectrum was obtained by a single passage of focused pump and probe beams through the optogalvanic lamp. By contrast, under such conditions, the absorption transitions were scarcely observable even with the laser beam transmitted four times through the sample. To render these transitions visible for measuring the sample transmittances it was necessary to use the lasers in the c.w. mode rather than as picosecond mode-locked lasers. For such conditions using the three-plate birefringent filter used for wavelength selection the dye lasers had a line-width of 0.3 - 0.4 cm^{-1}, giving the absorption spectrum displayed in Figure 2A when the laser was doubly-passed through the opto-galvanic lamp. Although these transitions in the direct absorption spectrum appear to be suitably strong for accurate absorbance measurements it is worth recalling that they are obtained with lasers which still have a line-width ~ 5 times greater than the linewidths of the transitions in the discharge (~ 0.08 cm^{-1}). Further, they are recorded with a discharge current of ~ 10 mA and since it was desirable to measure the species concentration over the greatest possible range of discharge currents (0.03 - 10 mA) the laser line width was narrowed further (to single frequency operation) using an intra-cavity etalon. With such a laser it was possible to measure the absorbances over this range of current densities with the laser beam singly-passed through the lamp. To enhance the accuracy of these measurements no attempt was made to measure the beam transmittances on and off frequency-resonance, but rather the transmitted single-frequency laser powers at frequency-resonance with the transitions were recorded by switching the discharge on and off.

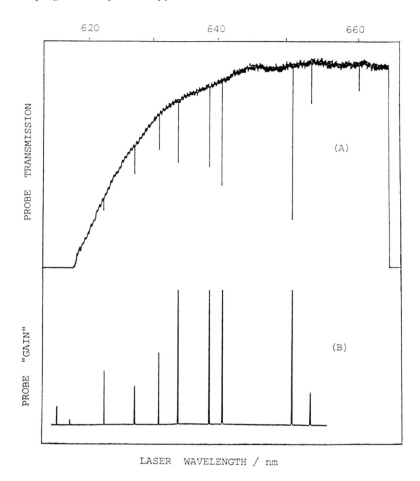

Figure 2. Absorption and COMAS "gain" spectra of the Hollow Cathode lamp at a discharge current of 10 mA.
(A) Transmission of the continuous-wave DCM laser doubly-passed through the lamp.
(B) "gain" spectrum of the focused picosecond mode-locked DCM probe laser singly-passed through the lamp.

The necessity of measuring the species absorption with continuous-wave in place of picosecond mode-locked lasers had repercussions for the COMAS measurements, which necessarily needed to be carried out with lasers identical with those employed for the absorbance determinations. Under normal circumstances, for concentration determinations such "gain" measurements would have been carried out with the single frequency laser using the continuous-wave methods described in our earlier publication (Mallawaarachchi et al. 1987). With this method the "gain" expression of equation (1) would have been replaced by the analogous expression

$$G = \frac{\pi\sigma^2 \cdot N\Delta \cdot i_p \cdot \tau}{\lambda}, \quad (4)$$

with i_p as the total number of photons incident per second (rather than the number of photons per pulse) and τ a lifetime term depending on the population kinetics of the levels involved in the transition. Combination of the absorbance and "gain" equations (equations (2) and (4)) would then have yielded an expression for the species concentration

$$N\Delta = \frac{\pi i_p \cdot \tau \cdot [\ln(I_o/I)]^2}{\ell^2 \lambda G}, \quad (5)$$

entirely analogous to equation (3) used for studies with pulsed lasers. By measuring the "gain" and the absorbance values for each of the transitions identified it would have been possible to have determined the absolute species concentrations in the vicinity of the focus for each of the lower states involved in the relevant transitions. In practice, although this approach was satisfactory for the weaker transitions, it was not suitable for measuring the stronger transitions because of difficulties in measuring the very large "gain" values created when single-frequency lasers were employed. Accordingly for these quantitative studies it was found preferable to employ unfocused pump and probe beams with the wavelength term λ in equations (4) and (5) replaced by A/ℓ, where A is the effective cross sectional area of the laser beams and ℓ is the effective length of the discharge. This modification to the COMAS method meant that in order to measure the absolute species concentrations not only was it necessary to measure the lifetime term τ of equation (5) but also the effective absorbing length ℓ and the cross sectional area A. For determining the latter it was necessary only to measure the radial power distribution within the pump and probe beams since, fortunately, the cross sectional area of the beams is very much less than the cross sectional area of the discharge. For measuring the effective length of the discharge, however, a very much more detailed consideration of the concentration distribution of the excited state species was necessary and this aspect of the species concentration measurement will be considered at this stage before discussing methods of determining the lifetime term τ of equation (5).

3.1 Species Concentration Profiles

Within the active region of the optogalvanic lamp discharge it is known that there occurs a non uniform distribution of excited state species concentrations both radially and longitudinally. Because of the manner in which the COMAS "gain" signal with focussed beams is created in the vicinity of the beam focus, and the fact that from equations (1) and (4) it is clear that the measured "gain" relates directly to the species concentration, it was possible to monitor the relative concentration profiles in 3-dimensions by moving the common focus of the pump and probe beams through the active zone of the discharge lamp. For convenience experimentally in our studies it proved easier to move the lamp in relation to the common beam focus and to scan the lamp position both radially and longitudinally using an appropriate linear-displacement motorized mount. Some typical examples of the resulting radial concentration profiles for various discharge currents are shown in Figure 3.

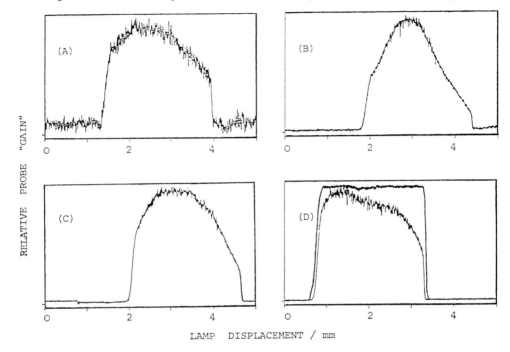

Figure 3. Radial "gain" variations across the hollow cathode lamp for various discharge currents.
(A) - 0.15 mA; (B) - 0.25 mA; (C) - 1.5 mA; (D) - 4.0 mA
Transition $2p_7 - 1s_3$ at a wavelength of 653.5 nm. The superimposed trace in (D) is the probe transmission through the lamp with the discharge extinguished.

From our analysis it is clear that there are four levels that exhibit a measurable species concentration. These are the four lowest excited state levels ($1s_2 - 1s_5$) in the term value diagram arising from the electron configuration $1s^2 2s^2 2p^5 3s^1$, transitions from these levels occurring (primarily) to the ten levels arising from the excited state electron configuration $1s^2 2s^2 2p^5 3p^1$. The particular set of profiles displayed in Figure 3 provides a measure of the radial concentration profiles of the $1s_3$ level (at a term value of 134,820.6 cm^{-1}) as determined using the transition ($2p_7 - 1s_3$) and a laser wavelength (in vacuum) of 653.5 nm. With a spot radius of ~ 15 μm for the focused beams the resolution of the method using collinear focused beams is more than adequate to provide an accurate measure of the concentration profiles so that the traces recorded in Figure 3 are thought to provide an accurate representation of the species distribution at the centre of the lamp.

Some of the radial profiles measured across the lamp shown in Figures 3(A) to 3(D) represent gradually increasing lamp currents. All of these profiles were scaled to the same height with the signals increasing generally from 0.15 mA (A) to 4.0 mA (D). With decreasing lamp currents below ~ 0.2 mA, under which conditions only a weak glow discharge is maintained, the profile is approaching a "top hat" shape, the profile becoming more strongly peaked toward the axis of the lamp for currents greater than ~ 0.2 mA before eventually flattening again with a near-"top hat" profile for the higher discharge currents (> 4 mA). With the focus located at the peak maximum it is found that for this transition the "gain" increases to a maximum at ~ 1 mA and thereafter decreases slowly. Under these conditions it would appear that with increasing current the species concentration is approaching a limiting value, the apparent reduction in the measured "gain" beyond 1 mA arising as a result of increasing attenuation of pump and probe beams as the species concentration increases slowly toward saturation. These factors are taken into account in our measurements of the species concentrations.

Examination of the radial distribution profiles (Figure 3) shows an asymmetry across the lamp which is not present in the profile of the transmitted laser power (Figure 3D), so that the asymmetry in the "gain" profile is clearly a true measure of the asymmetry of the discharge profile rather than a consequence of any artefact such as a variation in the transmittance of the lasers across the lamp. This asymmetry is presumably caused by the relative orientations of cathode and anode, the anode being set at a slight angle to the cathode and is therefore closer to the cathode on one side than the other resulting in a slight asymmetry in the current flow through the lamp.

The radial profiles of Figures 3(A) - 3(D) show clear evidence of a sudden change in the nature of the discharge near 0.2 mA, below this value the glow discharge exhibiting a "top hat" type profile, the discharge exhibiting a more gaussian type distribution just above this limiting current before again becoming "top hat" in type as the species concentration saturates at the peak and towards the wings of the gaussian distribution. Further evidence for this sudden discontinuity in the nature of the discharge is shown in Figure 6 which displays the variation of the $1s_4$ species concentration as a function of the discharge current. This discontinuity near 0.17 mA for our optogalvanic lamp occurs for all of the measured species ($1s_5 - 1s_2$) and reflects an identifiable change in the nature of the discharge rather than a discontinuity reflecting the $1s_4$ species concentration alone.

A further interesting feature of these radial intensity profiles is the suddenness of the intensity decrease near the walls of the hollow cathode. Although it is not possible for the method to monitor species concentrations right up to the wall surface, the transmittance profile of Figure 3D provides a measure of the "aperturing" of the beam by the hollow cathode and suggests that we are able to measure the variation of the species concentration to within a few "tens of microns" from the wall.

Ideally the longitudinal gain profiles should also be determined with collinear pump and probe beams. In practice, however, because of the length and the restrictive aperture (~ 2.7 mm diameter) of the lamp the spatial resolution of this approach, determined by a confocal parameter of ~ 2 mm, is not quite adequate. In order to obtain improved resolution, therefore, pump and probe beams were separated by 4.3 mm at the focusing lens in order to provide a pair of pump and probe beams intersecting at the focus, giving much better spatial resolution at a cost of a factor of ~ 10 degradation in the signal to noise ratio and a somewhat more restricted length scan because of the aperturing effect of the ~ 2.7 mm discharge bore. The longitudinal scans of the "gain" profile, which relate directly to the species concentration on the discharge axis, are shown in Figure 4, the upper trace (A) showing the profile for separated pump and probe beams, the lower trace (B) for collinear beams. The abscissa zero on each trace corresponds to the edge of the cathode nearest to the anode.

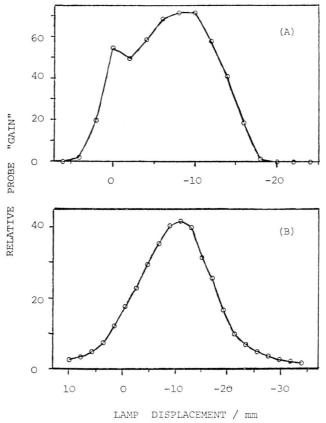

Figure 4. Longitudinal "gain" variations through the hollow cathode lamp at a discharge current of 5 mA.
(A) Crossed pump and probe beams (B) Collinear beams
Transition $2p_4 - 1s_4$ at a wavelength of 609.8 nm.

The results for the separated beams clearly give better resolution with more of the discharge structure being displayed. In the upper trace two distinct maxima are seen, the larger being positioned centrally within the lamp and the smaller in the region between anode and cathode where the plasma is not confined by either electrode and appears to increase in diameter. In the collinear case only one maximum is noted: the rate of decay of the plasma intensity on either side of this maximum is significantly slower, confirming that the resolution in this case is indeed inferior to the separated beam condition.

Since it was intended that all the quantitative measurements to be recorded on this lamp should be carried out with collinear beams, we fitted this profile to a suitable mathematical function to facilitate the determination of the absolute species concentrations. By far the most satisfactory profile was found to be a gaussian distribution represented by

$$n(Z) = n(O) \cdot \exp[-Z^2/Z_o^2] \ , \tag{6}$$

where $n(Z)$ is the appropriate species concentration at distance Z from the position of maximum plasma intensity. Z_o is a length determined as 10.5 mm for a discharge current of 10 mA and 10.0 mm for the 5 mA discharge.

From these radial and longitudinal discharge profiles and a knowledge of the absolute species concentrations at the positions of their maximum intensity it is clearly possible to obtain the species concentration at any point within the lamp for all the various discharge currents investigated. Following a discussion of the methods of measuring the various state lifetimes, the remainder of this paper will be concerned with the measurements of the species concentrations at this maximum intensity position.

3.2 Lifetime Measurements

One important aspect of concentration measurement in neon arises because the four lowest states being measured have relatively long lifetimes ($\lesssim$ 120 ns). Since the separation between consecutive pulses in the train of mode-locked pulses is only 12.2 ns it is clear that under these circumstances many preceding pulses contribute to the "gain" at any time and these contributions need to be taken into account to obtain a proper measure of the species concentration. This situation is closely analogous to the use of continuous-wave lasers for which the "gain" expression of equation (1) is modified to that of equation (4). For the simplest two level case where level 1 is the ground state of the system and level 2 is the excited state (lifetime τ_2), the rate of repopulation of level 1 equates to the rate of decay of level 2 and $\tau = 2\tau_2$. This is not the case for the states involved in the transitions in the opto-galvanic lamp and levels 1 and 2 are repopulated separately from the discharge as well as from other levels in the system at rates k_1 ($= 1/\tau_1$) and k_2 ($= 1/\tau_2$), respectively. Two separate signal contributions arise, one from the "hole" created in the N_1 concentration via probe absorption and the other via stimulated emission of the probe from the perturbed concentration created in level 2. For this condition τ in equation (4) is replaced by $(\tau_1 + \tau_2)$, both of which must be known in order to determine the species concentration N_1.

The usual way in which state lifetimes are measured with mode-locked lasers is to scan the probe pulse in relation to the pump from pre- to post-pulse-coincidence. Used with the COMAS (one wavelength) arrangement this gives a decay curve which depends on the lifetimes of the upper and lower states. Analysis of such a curve is not always easy, particularly when upper and lower states have rather similar lifetimes. An approach which avoids this difficulty and simplifies the lifetime analysis is to use two separate wavelengths coupled to two different transitions but having one level, either the upper or the lower, in common. If it is the lower level, for example, the pump coupled to one transition depletes the population of the lower level which is monitored in time by the probe laser coupled to a different transition. The resultant temporal decay curve depends only on the lifetime of the lower level and acquires a comparatively simple exponential decay form. A typical example (here for a common upper level) is shown in Figure 5, this particular curve showing the lifetime decay curve for the $2p_4$ level at 150, 860.5 cm^{-1}.

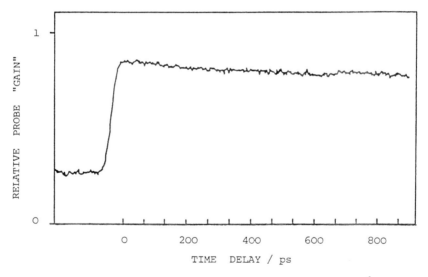

Figure 5. Temporal variation of the "gain" for the $2p_4$ level at 150,860.5 cm^{-1} for a discharge current of 0.19 mA.
Pump laser: $2p_4 - 1s_4$ at 609.8 nm; Probe laser: $2p_4 - 1s_2$ at 668.0 nm.

From this diagram it is clear that there is a large residual gain from preceding pulses just before pulse coincidence. The ratio of the pre-coincidence to post-coincidence "gain" for not-too-high values of this ratio (< 0.5) is an excellent way of measuring the lifetime for a simple exponential decay since the pulse to pulse separation is known very precisely - from the ratio given by figure 5 we measure the lifetime of the $2p_4$ state as 18 ns. Although this is not the ideal way of measuring these state lifetimes, it provides a reliable method for lifetimes in the range 10-50 ns. The difficulties of measuring the lifetimes in neon arise not for these relatively short lived species but for those species, such as $1s_5$ and $1s_4$ which have very much longer lifetimes. Our measurements gave for the four lowest states the following lifetimes: 120 ns ($1s_5$), 120 ns ($1s_4$), 26 ns ($1s_3$) and 14 ns ($1s_2$). We are aware that these lifetimes are considerably shorter than typical wall loss times for glow discharges (Klose 1966; Phelps 1959).

However, the lifetimes measured here correspond to the total destruction rates of the species, not just the diffusion or resonance radiation loss rates. Although we might have expected the $1s_3$ and $1s_5$ lifetimes to be approximately equal, our measured values differ significantly. The measurement of the $1s_3$ lifetime, however, does not compare directly with the other three measurements because for this state it did not prove possible to find transitions in the ranges of the Rh.6G and DCM lasers with this as the common lower level. For $1s_3$ it was necessary to use pump and probe lasers of the same wavelength, thereby involving upper and lower state lifetime processes in the decay curve and increasing the uncertainty of the measurement.

The lifetimes of $1s_5$ - $1s_2$ enter directly into the expressions for the species concentrations and separate independent measurements would, undoubtedly, give added confidence to the concentrations determined in this study. Improved values for these lifetimes could readily be incorporated into these calculations and thence into the measured values for the various absorption cross sections from these states. For present purposes, however, it is appropriate to include the τ values, albeit with some reservations, evaluated directly in this study.

4. ABSOLUTE CONCENTRATION MEASUREMENTS

With the measured lifetimes and the longitudinal variation of relative species concentrations it was possible to proceed to the evaluation of the absolute species concentrations in the four lowest excited states. As mentioned previously, the "gain" values were determined for unfocused pump and probe beams whose radial power distributions were measured by scanning a 100 μm diameter aperture across the beams. From the resultant laser beam profiles and the measured value of the gaussian longitudinal discharge profile the species concentrations at the centre of the discharge were evaluated for currents in the range 0.03-8 mA. These results for the $1s_4$ state at 134,461.3 cm^{-1} are plotted in graphical form in Figure 6, the data being divided into two scales, the upper diagram (6A) emphasizing the low current regime displaying the significant change in the nature of the discharge near 0.17 mA, the lower diagram showing the complete current range and the approach to saturation at higher currents. Similar curves showing the changing nature of the discharge at the lowest currents and the curvature reflecting the approach to saturation were also found for the $1s_5$ and $1s_3$ states. By contrast the $1s_2$ level has a very much lower concentration which increases linearly with increasing discharge current with no trace of the saturation that characterizes the three lower levels.

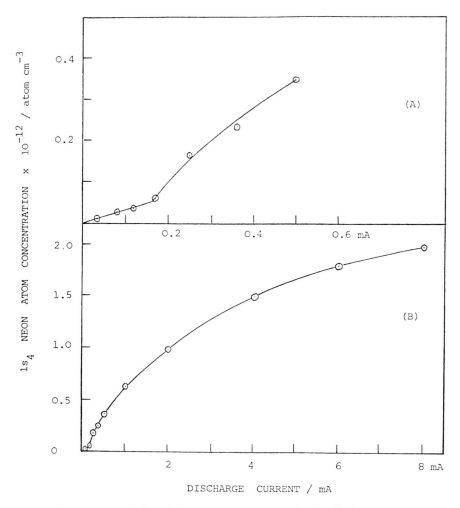

Figure 6. The variation of the $1s_4$ neon atom concentration with discharge current. (A) The low current regime; (B) The full current range of the hollow cathode discharge.

From the concentration/current variations for the four lowest states a number of clear deductions are possible. At the very lowest currents for a sustainable glow discharge ($\simeq 0.03$ mA) the only level which exhibits any significant population is the $1s_5$ level, the population of this level ($\sim 1.5 \times 10^{11}$ atoms cm^{-3}) scarcely changing in the current range 0.03-0.17 mA. In this same current regime populations of the $1s_4$ and $1s_3$ levels are initially more than an order of magnitude smaller but rise rapidly to become more comparable with $1s_5$ at ~ 0.17 mA. Beyond this current the nature of the discharge appears to change rapidly, presumably as the hollow cathode discharge is established and the population of $1s_5$ rises very rapidly with a rapid onset of saturation toward a maximum concentration of

~ 2.6×10^{12} atoms cm^{-3} as the current is further increased. $1s_4$ and $1s_3$ behave in a rather similar fashion, the latter ($1s_3$) rising more rapidly from a lower initial concentration but exhibiting a more rapid onset of saturation, both levels tending toward roughly the same saturation concentration exhibited by $1s_5$.

The data supplied by these concentration/current curves contain significant additional information on the kinetic processes leading to the population of the various levels studied. Assuming initially that the excited state lifetime is lamp current independent and that the rate of population of the excited states is linearly dependent on the discharge current then, simplistically, the rate equation may be written as

$$\frac{d[Ne^*]}{dt} = k_1 I [Ne] - k_{-1} [Ne^*] , \qquad (7)$$

where [Ne] refers to the concentration of ground state neon atoms, [Ne*] to the concentration of the excited state under question, I is the discharge current and k_1 and k_{-1} are the rate constants for excitation and de-excitation, respectively. At steady state $d[Ne^*]/dt = 0$ and the excited state concentration is given by

$$[Ne^*] = \frac{k_1}{k_{-1}} I [Ne] , \qquad (8)$$

suggesting a linear relationship between the excited state species concentration and the lamp current. Apart from the $1s_2$ state, for which the experimental data provides clear evidence for such a relationship, the curvature of the concentration/current curves for states $1s_5$ - $1s_3$ suggests a deviation from equation (8) at higher discharge currents. In order to account for this deviation a second de-excitation path linearly dependent on lamp current was introduced. This reaction, with rate constant k_2, can be included in the rate equation leading to an excited state concentration

$$[Ne^*] = \frac{k_1 I [Ne]}{k_{-1} + k_2 I} . \qquad (9)$$

Rearranging this equation suggests a linear relationship between the reciprocal of the excited state species concentration [Ne*] and the reciprocal of the discharge current I, with intercept of $k_2/k_1[Ne]$ and slope of $k_{-1}/k_1[Ne]$. That such an interpretation does indeed accord with the data is apparent from Figure 7 which displays in this rearranged form the data of the $1s_4$ level displayed in Figure 6. Only the data corresponding to discharge currents of 0.25 mA or greater are included, since for milder conditions the hollow cathode discharge does not appear to be fully established. Because of the rapid onset of saturation for the $1s_5$ state there is significantly more scatter for this plot of the inverted form of equation (9) but the same general trend is apparent. For the remaining two states ($1s_3$ and $1s_2$) there is significant deviation from the expression proposed at low discharge currents (< 1 mA), for which the species concentration is significantly lower than predicted, probably due to the incomplete attainment of the hollow cathode condition at these currents for these particular states.

From the slope and intercept of Figure 7 a "least squares" fit gives $k_3/k_1[Ne] = 1.427 \times 10^{-12}$ mA.cm^{-1} atom^{-1} and $k_2/k_1[Ne] = 0.276 \times 10^{-12}$ cm^3 atom^{-1}, respectively. From the ground state neon atom concentration and taking k_{-1} as 8.33×10^6 s^{-1}, the rate constants k_1 and k_2 for excitation from the ground electronic state and from $1s_4$ are found to be, respectively, 27.5 mA^{-1} s^{-1} and 1.61×10^6 mA^{-1} s^{-1}. That k_1 is very much less than k_2, the equivalent rate constant for the current induced de-excitation of $1s_4$, is understandable since this excited state is ~ 135,000 cm^{-1} above the ground state but only ~ 40,000 cm^{-1} below the ionization limit so that much more energetic electrons are required for the excitation process governed by k_1 than for the process governed by k_2. Similar rate constants are derived for the excitation and the de-excitation of the other three excited states ($1s_2$, $1s_3$ and $1s_2$), and the concentration measurements and the saturation concentrations can be rationalized in terms of these rate constants and the lifetimes of the relevant states.

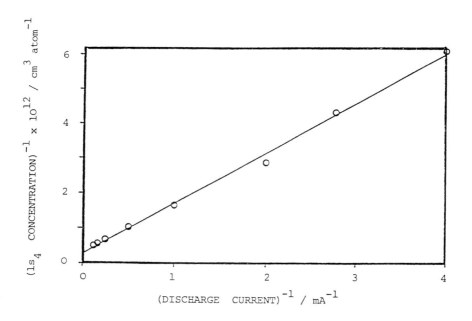

Figure 7. The variation of the reciprocal of the $1s_4$ neon atom concentration with the reciprocal of the discharge current over the range 0.25 - 8 mA.

In conclusion it is worth emphasizing that although we have confidence in the species concentration distributions in the discharge we retain some reservations concerning the measured lifetimes of $1s_5$ and $1s_4$ because of the difficulty of measuring such long lifetimes using picosecond mode-locked lasers. Since these lifetimes enter directly into the absolute species concentration measurements in this method it is proposed that to confirm these values further studies be undertaken using continuous wave single-frequency lasers and the variable modulation frequency approach of our earlier publications. Notwithstanding such reservations, however, we feel that the benefits of this concentration-modulation approach to the study of absorption spectra, yielding information on excited state lifetimes, absolute species concentrations and thereby rate constants for excitation and de-excitation processes, are substantial and will be of significant value for the study of the kinetic processes that are of such importance in discharge devices.

Acknowledgements

We are indebted to the Science and Engineering Research Council for research studentships (to TRG and GS) and for the grants which made the original purchase of the equipment employed possible. We are also deeply indebted to the many support staff of the Departments of Chemistry at Aberystwyth and Swansea, whose contributions made this work possible.

References

Beaman R A, Davies A N, Langley A J, Jones W J and Baran J 1986 *Chemical Physics* **101** 127
Jones W J 1987 *J. Chem. Soc., Faraday Trans* **83** 693
Klose J Z 1966 *Physical Review* **141** 181
Langley A J, Beaman R A, Baran J, Davies A N and Jones W J 1985 *Optics Letters* **10** 327
Langley A J, Beaman R A, Davies A N, Jones W J and Baran J 1986 *Chemical Physics* **101** 117
Mallawaarachchi W, Davies A N, Beaman R A, Langley A J and Jones W J 1987 *J. Chem. Soc., Faraday Trans.* **83** 707
Czarnik-Matusewicz B, Griffiths R, Jones P F, Jones W J, Mallawaarachchi W and Smith G 1988 *J. Chem. Soc., Faraday Trans.* **84** 1867
Phelps A V 1959 *Physical Review* **114** 1011.

Measurement of the copper vapour concentration in a pseudospark discharge by laser-induced fluorescence

Günter Lins

Siemens AG, Corporate Research and Development, ZFE ME TPH 32, P.O. Box 32 20, 8520 Erlangen, Federal Republic of Germany

ABSTRACT: The copper vapour density produced by a pseudospark driven by a rectangular current pulse of 200 ns duration and 900 A amplitude was measured by laser-induced fluorescence. Copper vapour was observed from 2 µs till 8 ms after initiation of the discharge. The long residence time cannot be explained by collisionless expansion or diffusion, but indicates the presence of metal droplets evaporating in flight. The maximum density is of the order of 10^{19} m^{-3}. Pseudosparks with higher current and longer duration are expected to produce so much vapour that they eventually turn into a metal vapour arc.

1. INTRODUCTION

A pseudospark is a low-pressure, high-current discharge which was first described by Christiansen and Schultheiß (1979). Pseudosparks can be sources of beams of electrons and ions, X-rays and microwaves and have been used in high-power switching technology and as the active medium in an argon ion laser (Frank and Christiansen 1989).

A simple pseudospark device as shown in Figure 1 basically consists of two circular plane parallel plates kept at a separation d by an insulator. Both electrodes have a concentric hole with a diameter of the order of the electrode separation. The pressure p of the ambient gas is chosen such that the product p*d corresponds to a point on the left-hand-side of the self-breakdown curve of the system, i.e. increasing p*d results in a lower breakdown voltage. If the device is overvolted or triggered, ignition will occur along the longest available path, i.e. in the region of the electrode holes.

Although the cathode is initially cold, pseudosparks are capable of operating in a glow mode at current densities of the order of 10^4 A/cm^2 (Braun et al 1988, and Hartmann et al 1988) without significant cathode erosion.

© 1991 IOP Publishing Ltd

The emission mechanism underlying this remarkable property is not yet fully understood. In a recent model by Hartmann and Gundersen (1988) it is proposed that thermionic emission caused by ion beam heating of the cathode plays an essential role in the conduction mechanism. The cathode temperature of some thousands of Kelvins required by the model should result in strong evaporation of cathode material. Hence a measurement of the concentration of evaporated cathode material and the time when it appears in the electrode gap should be of great interest for understanding this type of discharge.

2. EXPERIMENT

The cathode of the pseudospark device shown in Figure 1 was 59 mm in diameter and made of copper, the anode consisted of aluminium. In order to enable optical access from all sides, a quartz glass tube with 60 mm inner diameter was used as the insulator. Electrode separation and hole diameter were 5 mm, the working gas was chosen to be nitrogen. The experiments were carried out at a pressure of 9 Pa and a voltage of 18 kV, i.e. 1 kV below the self-breakdown voltage of the system.

To produce rectangular current pulses five coaxial cables of 50 ohms surge impedance and 20 m length were charged to 18 kV and discharged into five 50 ohm resistors via the pseudospark device as indicated in Figure 2. The resultant current pulse had an amplitude of 900 A, a rise time of 20 ns and a duration of 200 ns (FWHM). The current was measured using a shunt of 0.01 ohms.

The pseudospark was triggered by the light of a flashlamp-pumped dye laser which was focused onto the rear side of the cathode through a quartz window (Figure 1) thus generating a laser-produced plasma which initiated the discharge (Chuaqui et al 1989). The focus was about 0.5 mm in diameter, the laser

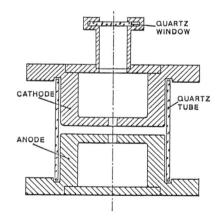

Fig. 1. Experimental pseudospark device

energy measured in front of the quartz window amounted to 350 mJ. Pulse duration and wavelength were 350 ns and 650 nm, respectively. The contribution of copper vapour released by laser ablation to the total vapour density was found to be less than 5 percent.

The laser-induced fluorescence of copper was described in some detail by Lins (1985). The experimental setup is shown in Figure 3. The frequency-doubled light of a Nd:YAG laser was used to pump a three stage dye laser which was operated at 649.5 nm. The resulting radiation was frequency-doubled to 324.8 nm, the wavelength of one of the resonance lines of copper, and subsequently directed into the interelectrode space with the optical axis 2.5 mm above the cathode. Two crossed slits, 2 mm wide, determined the cross section of the beam. The energy per pulse amounted to 2 mJ behind the slits, the pulse duration being 6 ns. For the bandwidth of 15 pm the saturation parameter for the intensity of 8 MW/cm^2 thus generated amounted to 40000 in the maximum of the laser pulse. To detect the fluorescence radiation at 510.6 nm a combination of a photomultiplier with a quarter metre monochromator was used. The optics were arranged such that the observation volume was a cylinder of 2 mm length and 1 mm diameter. The detection scheme was calibrated using Rayleigh scattering in SF_6.

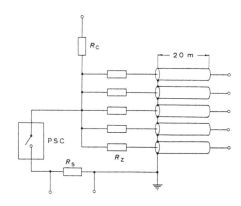

Fig. 2. Pulse-forming network for generation of near-rectangular current pulses. PSC: pseudospark chamber, R_c: charging resistor, R_s: shunt resistor, R_z: 50 Ohm load resistor

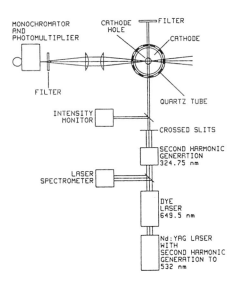

Fig. 3. Experimental setup for LIF experiment

3. RESULTS AND DISCUSSION

The sequence of events is outlined in Figure 4. About 400 ns after initiation of the trigger pulse, the pseudospark was ignited, producing a near-rectangular current pulse. As the current started flowing, strong light emission by the pseudospark plasma was observed which disappeared only 2 µs after triggering.

The results of the fluorescence measurements are displayed in Figure 5. Each point represents an average of 10 to 15 discharges; error bars give the highest and lowest density obtained at the respective time. A measurable amount of copper vapour appears at about two microseconds after triggering. The density rises steeply to a maximum value of $2.6*10^{18}$ m^{-3}. Then it drops to 10^{17} m^{-3} within 30 µs. This is followed by another increase to $2.5*10^{18}$ m^{-3} and a subsequent very slow decay, such that copper vapour is found to exist in the electrode gap even 8 ms after the end of the discharge.

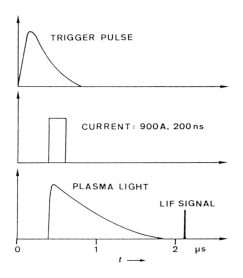

Fig. 4. Sequence of events

The drop in density after the first maximum is interpreted as being due to absorption of the incident laser radiation by the vapour which becomes optically thick at a sufficient density. As a consequence, saturation is no longer achieved, and the density inferred from the fluorescence intensity is too low.

For a drastic reduction of the fluorescence signal to occur, the incoming laser intensity of 8 MW/cm^2 has to be attenuated by a factor of about 10^{-6}. If an optical path length of 1 cm and a vapour temperature of T = 1356 K (the melting temperature of Cu) are assumed, a rough estimate using the absorption coefficient in the centre of a Doppler-broadened line according to

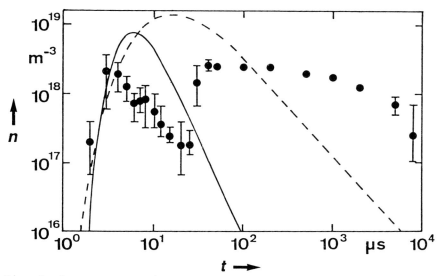

Fig. 5. Copper vapour density n as a function of time. Solid line: vapour density calculated from a model of collisionless expansion into vacuum. Broken line: vapour density from a diffusion model.

Mitchell and Zemansky (1971) reveals that the copper vapour density should be in excess of $3.7*10^{18}$ m^{-3}.

3.1 Vapour Propagation by Collisionless Expansion into Vacuum

If the vapour atoms spread out by collisionless expansion into vacuum, the instantaneous release of a number of N atoms with a Maxwellian velocity distribution from a point source results in a density n at distance r and time t given by (Lins 1987):

$$n(r,t) = (\frac{m}{2\pi kT})^{3/2} \frac{N}{t^3} exp(-\frac{m}{2kT}\frac{r^2}{t^2}) \qquad (1)$$

where m is the atomic mass, k Boltzmann's constant, and T the temperature. The vapour density at a given point in space is obtained if (1) is integrated over that area A of the cathode which is touched by the discharge plasma (Lins 1987). In doing so it is assumed that the vapour production per unit area can be expressed by N/A. Following the erosion pattern found on the cathode the surface for the integration was given the shape shown in Figure 6. The area of the surface thus defined amounts to A = 1 cm^2.

The number $N = 1.7*10^{13}$ of evaporated atoms was obtained assuming an erosion rate of 10 μg/C which was multiplied by the charge of $1.8 * 10^{-4}$ C transferred during a single discharge. The actual erosion rate is not known for this type of discharge. In the vacuum arc literature erosion rates ranging from 50 to 190 μg/C are reported for copper (Farrall 1980). Since only a minor fraction of the eroded material is emitted as neutral metal vapour a smaller erosion rate appears to be justified in the description of the vapour production. Again the temperature was taken to be the melting temperature of copper, T = 1356 K. The result is shown in Figure 5. The transit time of the atoms from the cathode to the observation volume is described correctly. The maximum density amounts to $7.2*10^{18}$ m^{-3}.

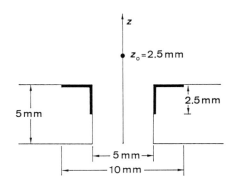

Fig. 6. Region in the vicinity of the cathode hole. The heavy lines denote the area over which equations (1) and (2) were integrated. The vapour density was measured at z_0 = 2.5 mm.

The assumption of an instantaneously acting vapour source is a good approximation so long as the duration of the discharge is short as compared to the time t for which n is calculated. In the present case this is certainly true for t>2 μs. The generalization to finite duration of the source activity is straightforward as shown by Lins (1987).

3.2 Vapour Propagation by Binary Diffusion

The long-lasting presence of the vapour is not reproduced correctly by the model of collisionless expansion. Since for the pressure of 9 Pa of the filling gas the mean free path for collisions of copper atoms with nitrogen molecules is about 0.8 mm, the propagation of cathode vapour was also calculated on the basis of diffusion. According to Crank (1967) the density n(r,t) caused by atoms propagating by diffusion from an instantaneous point source located at r = 0 is given by the following equation:

$$n(r,t) = \frac{N}{8(\pi Dt)^{3/2}} exp(-\frac{r^2}{4Dt}) \qquad (2)$$

where D is the binary diffusion coefficient. Again this was integrated over the surface shown in Figure 6, using essentially the same data as in the calculation based on collisionless expansion. The diffusion coefficient D was calculated according to Waldmann (1958) treating atoms and molecules as hard spheres:

$$D = \frac{3}{8Cd_{12}^2}(\frac{kT}{2\pi m_{12}})^{1/2} \qquad (3)$$

m_{12} being the reduced mass, $d_{12} = (d_1 + d_2)/2$ and, C the concentration of molecules. The diameters of copper atoms and nitrogen molecules, $d_1 = 3.46 * 10^{-10}$ m and $d_2 = 3.5 * 10^{-10}$ m, were taken from Gmelin (1955) and Waldmann (1958), respectively. The temperature of the nitrogen was assumed to be 300 K. The result is represented by the broken line in Figure 5. Again the transit time is well reproduced. The maximum density is slightly higher than in the case of collisionless expansion. Although the residence time of the Cu atoms in the electrode gap is considerably prolonged the measured density still decays more slowly than the calculated one. A similar behaviour is known from vacuum arcs and has been attributed to the presence of metal droplets evaporating in flight (Jenkins et al 1975).

4. CONCLUSIONS

The copper vapour density produced by a pseudospark of short duration (200 ns) and moderate current (900 A) operating on a copper cathode was measured using laser-induced fluorescence. It was found that two microseconds after initiation of the discharge the vapour density rises to such a value that the incident laser radiation is strongly absorbed. Estimations based on models of free expansion into vacuum or diffusion of copper vapour in nitrogen indicate that depending on the rate of erosion assumed vapour densities between 10^{19} and 10^{20} m^{-3} can be reached. For higher current and longer pulse duration it is expected that the vapour pressure wil come close to the pressure of the filling gas and that the discharge eventually will turn into a metal vapour arc. At high repetition rates, the properties of the pseudospark discharge will be influenced by the residual vapour left by the preceding discharge.

Finally it should be noted that copper was used as a cathode material due to the great deal of experience with LIF of copper. In pseudospark devices refractory metals such as molybdenum and tungsten are generally preferred. These may yield results very different from those reported here.

ACKNOWLEDGEMENTS

The author would like to thank Prof J Christiansen and Drs K Frank and W Hartmann for their constructive advice on the operation of the pseudospark.

REFERENCES

Braun C, Hartmann W, Dominic V, Kirkman G, Gundersen M and McDuff G 1988 *IEEE Trans. Electron. Devices* **35** 559-562
Christiansen J and Schultheiß Ch 1979 *Z. Physik* **A 290** 35-41
Chuaqui H, Favre M, Wyndham E, Arroyo L and Choi P 1989 *Appl. Phys. Lett.* **55** 1065-1067
Crank J 1967 *The Mathematics of Diffusion* (Oxford) p 27
Farrall G A 1980 *Current Zero Phenomena* in Lafferty J M (Ed.) *Vacuum Arcs, Theory and Application* (New York: Wiley) pp 184-227
Frank K and Christiansen J 1989 *IEEE Trans. Plasma Sci.* **17** 748-753
Gmelins Handbuch der Anorganischen Chemie: Kupfer 1955 (Weinheim, Germany: Verlag Chemie) part A, vol 2, p 968
Hartmann W and Gundersen M A 1988 *Phys. Rev. Lett.* **60** 2371-2374
Hartmann W, Dominic V, Kirkmann G F and Gundersen M A 1988 *Appl. Phys. Lett.* **53** 1699-1701
Jenkins J E, Sherman J C, Webster R and Holmes R 1975 *J. Phys. D: Appl. Phys.* **8** L139-L149
Lins G 1985 *IEEE Trans. Plasma Sci.* **PS-13** 577-581
Lins G 1987 *IEEE Trans. Plasma Sci.* **PS-15** 552-556
Mitchell A C G and Zemansky M W 1971 *Resonance Radiation and Excited Atoms* (Cambridge) p 116
Waldmann L 1958 *Transporterscheinungen in Gasen von mittlerem Druck* in S Flügge (Ed.) *Encyclopedia of Physics*, vol XII, *Thermodynamics of Gases* (Berlin: Springer) p 427

Laser-induced breakdown spectroscopy as an analytical tool

B J Goddard, R M Allott and H H Telle

Dept of Physics, U C Swansea, Singleton Park, Swansea SA2 8PP

ABSTRACT: The technique of laser–induced breakdown spectroscopy offers a fast, simple method of elemental analysis without the need for sample preparation. We describe a series of experiments under vacuum conditions in which trace elements were detected in the bulk and as surface contaminants. Furthermore, parameters of the laser–produced plasma were deduced which are of importance for quantitative investigations; for example, plasma temperatures were derived and the temporal evolution of the plasma was investigated. Representative examples for a number of targets are given.

1. INTRODUCTION

For many techniques of optical spectral analysis the composition of a sample, in terms of the concentrations of constituent elements, is determined by the intensity distribution in the optical line emission spectrum. However, this is only possible if a representative portion of the sample can be converted into vapour and heated to a sufficiently high temperature so that a significant number of excited states are populated. Conventional thermal energy sources are electrical discharges, arcs and sparks, and plasmas of various forms. In most cases all of these methods require extensive sample preparation; for example, an insulator must somehow be mixed with a conductor to produce an electrode suitable for an electrical arc source. In general, therefore, these methods are not suitable for *in–situ* or local analysis, which is essential in some applications, such as remote analysis of structural material (for example, in or near nuclear reactor cores (Tozer 1976)), or for on–line analysis for example, direct analysis in steel smelting (Krupp 1990)).

High powered lasers provide an ideal source of thermal energy for vapourization. The low divergence of laser beams, in the order of milliradians, means that a very small spot can be obtained by focusing, and if the laser power density is high enough a small amount of material can be vapourized, irrespective of the conducting properties of the sample. The laser intensity may be high enough that the atoms in the vapour are excited or ionized to a high degree by the processes of multi–photon ionization (MPI) and inverse Bremsstrahlung absorption, permitting spectroscopic study of the resulting plasma emission. A number of reviews of the principles and applications of laser microspectral analysis have been published (see e.g. Adrain (1980) and Laqua (1985)); more recently the optimum conditions for laser ablation and atomization for analytical purposes, using an argon buffer gas at pressures between 1 mbar and 1000 mbar, were investigated by Leis et al (1989).

In this paper we describe the experimental arrangement used to perform laser–induced breakdown spectroscopy (LIBS), both on conducting and dielectric materials, in a vacuum environment. Spectra are discussed for some representative targets with a view to quantitative analysis of bulk materials and surfaces, including estimates of plasma temperatures and investigation of the time evolution of the plasma.

2. EXPERIMENTAL

The experimental apparatus used for the LIBS work is shown schematically in figure 1. The 532 nm frequency–doubled output of a Nd:YAG laser (JK Lasers HY 750), operating at 20 Hz, was focused by a simple lens ($f = 300$ mm) onto the surface of the target, contained within a stainless steel vacuum chamber. A 200 litre/sec turbomolecular pump (Edwards EXT 200) was used to maintain a vacuum of around 10^{-6} mbar during the sputtering process. A fraction of the isotropically emitted radiation was directed onto the entrance slit of a monochromator (Bentham M300HR) via quartz lenses and a single large–core optical fibre. The monochromator had a focal length of 300 mm and a reciprocal dispersion of 2.7 nm/mm, and the widths of the entrance and exit slits were continuously variable between 10 μm and 5 mm, allowing the spectral resolution to be adjusted as necessary.

After dispersion the light entered a photomultiplier (Thorn EMI 9816 QB), and the resulting signal was applied to the input of a pre–amplifier mounted within the photomultiplier housing. The output from this amplifier provided the input to a gated integrator/averager Boxcar unit (Stanford SR250), and was also directed to an oscilloscope to provide a real–time display. The laser power was monitored by a combination of pulsed–energy meter (Molectron J3–05) and purpose–built SRS compatible Boxcar type integrator/averager unit. The temporal reference for the experiment was provided by a PIN photodiode (BPX 65), positioned in a reflected portion of the Nd:YAG beam, in conjunction with a fast, variable threshold, pulse shaper/amplifier unit to ensure reproducible triggering. Overall control of the experiment was accomplished by an IBM compatible PC, which also provided the control signals to scan the monochromator and to rotate the target (mounted on a four–phase 0.9⁰ per step stepper motor), when necessary.

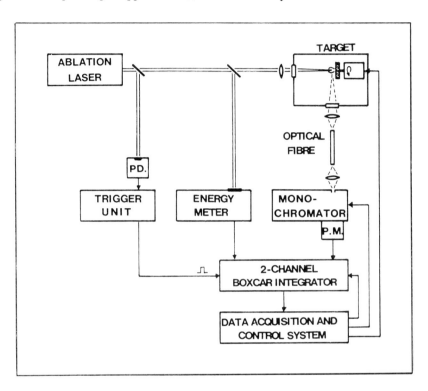

Figure 1. Experimental set–up for temporal and compositional analysis of samples using LIBS

For the results presented in this paper typical ablation laser pulse energies were 20 mJ, corresponding to a maximum power density at the target of 4 GW cm^{-2} (4 × 10^9 W cm^{-2}), assuming a focused spot diameter of the order of $f\Theta$, where Θ = 0.8 mrad is the divergence of the ablation laser beam. At these power densities it was observed that the intensity of the light emitted from the plasma diminished significantly after a certain number of laser shots, as a hole was drilled into the surface of the target. This was overcome by rotating the target by 0.9⁰ after an appropriate number of laser shots (typically 20–40), with the gated signal from the photomultiplier being averaged over this time for each data point. After the averaged value had been recorded by the PC the monochromator grating was advanced by a set amount, varying between 0.015 nm (1 step) and 0.225 nm (15 steps), and the next data point was recorded. A spectrum covering 400 nm therefore is obtained in around 1 hour, whilst at high resolution a short scan of 10 nm takes around 20 minutes. This is significantly longer than the time necessary to obtain spectra using a multi-channel analyser; however, this method exhibits the advantage of adjustable resolution, and furthermore allows easy time resolved scans of spectral lines. Time scans were accomplished by setting the monochromator to transmit the wavelength of interest and using the software-controlled digital delay generator incorporated in the Boxcar unit to scan the gate delay. The gate width was fixed at a value approximately equal to the delay increment between data points.

3. RESULTS AND DISCUSSION

For quantitative analysis the most important requirement is that the LTE approximation is valid at the particular observation time chosen. This allows the use of a Boltzmann distribution function of excited state populations to determine relative atomic concentrations from the observed intensity of spectral lines. In the following sections the validity of the LTE assumption is investigated by measuring the plasma temperature using Boltzmann plots for various targets, followed by a discussion of the effects on LTE of the plasma evolution by means of time-resolved spectra. Finally we describe some quantitative measurements, including trace element and surface contaminant detection, and constituent concentrations in alloys and compounds.

3.1 LTE Approximation and Plasma Temperature

In order to test the validity of the LTE approximation three targets with different macroscopic structure and ablation properties were investigated.

In a first example, sodium chloride was used as a target to investigate the applicability of the technique to dielectric samples, with a view to a possible extension of the experiment to include a surface ionization detector to measure the sputtered sodium neutral to ion ratio and the spatial distribution of the products (preliminary results are reported in Tambini and Telle to be published in Laser Chemistry). It was found that, because of the soft nature of the crystalline NaCl blocks, it was necessary to step the target every 30 laser shots, and reduce the laser pulse energy to 15 mJ. Even at this pulse energy the large amount of material ejected meant that the minimum operating pressure was 10^{-5} mbar. The wavelength range 300 nm to 600 nm was scanned at a gate delay of 200 ns, using

Plasma Diagnostics 245

a monochromator wavelength increment of 0.075 nm (to avoid the risk of damage to the photomultiplier the high voltage to the unit was switched off when the laser wavelength was reached). The spectrum obtained is shown in figure 2; over 100 peaks across the wavelength range were identified using tables of neutral and ionic lines (Striganov 1968). Spectra were also obtained at other gate delays.

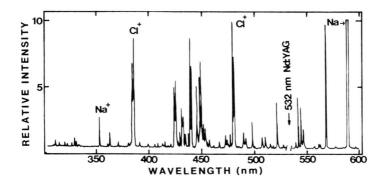

Figure 2. LIBS spectrum for NaCl target, observed 200 ns after excitation; some characteristic lines of neutral and ionic species are indicated

Boltzmann plots, shown in figure 3a, were constructed for the 200 ns and 400 ns delay data from the measured intensities, reference transition probabilities and degeneracies (A and g values) of a number of ClII lines (Handbook of Chemistry and Physics 1984). Plasma temperatures of 19300 K (1.7 eV) and 15300 K (1.3 eV) respectively were obtained from the gradients of the curves. As expected, the plasma temperature for the 400 ns data is lower than that for the 200 ns data, and, perhaps more significantly, the points obtained from the 400 ns data exhibit greater scatter around the straight line. This is an indication that at longer time delays the LTE approximation becomes less valid, suggesting that a compromise is necessary in the choice of gate delays for quantitative analysis; although low noise levels are obtained at longer gate delays the agreement with an LTE approximation is poorer. The fact that a straight line fits the data well at 200 ns delay means that the main energy transfer mechanism at this stage in the plasma evolution is collisional, rather than radiative transfer.

A knowledge of the plasma temperatures also allows an estimate for the plasma electron density n to be made; it can be obtained from a measurement of $\Delta\lambda_{1/2}$, the observed half-widths of selected spectral lines, since quadratic Stark broadening by charged particles in the plasma increases the observed linewidths. Although other

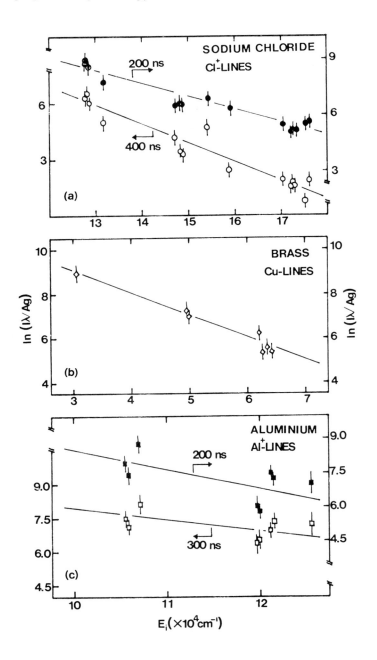

Figure 3. Boltzmann plots for LIBS spectra of (a) NaCl target, (b) Brass target and (c) Al target; for details see text

broadening mechanisms are present, such as Doppler and instrumental broadening, the value obtained from such a measurement will still provide an *upper* limit of the electron density. An approximate value for n, at which a given degree of quadratic Stark broadening by electrons occurs, can be obtained from reference tables (Griem 1964). In order to make a sensible estimate a transition was chosen for which the magnitude of the Stark broadening is very large for a given electron density, since this will reduce errors produced by the relatively large line width inherent in our instrument. From the reference values, the weak NaI line at 466.4 nm has a Stark broadening coefficient of 0.77 nm per 10^{16} electrons cm^{-3} at a plasma temperature of 20000 K. The FWHM of this line was measured from the NaCl spectrum and was found to be 0.75 nm, corresponding to an upper limit to the electron density of 9.7×10^{21} m^{-3}. The expansion velocity of the plasma particles is assumed to be thermal, of the order of 10^3 ms^{-1}, although there is some evidence to suggest that expansion velocities can in some cases be up to an order of magnitude higher than the thermal velocity (Ready 1971). A laser spot diameter of 300 μm with a typical crater depth of 2 μm implies the vapourization of $\simeq 10^{16}$ atoms of Na and Cl. If the assumption is made that $\simeq 0.01$ % of these are ionized (Schueler and Odom 1987), the number of electrons created is 10^{12}. After 200 ns the plasma radius will expand to $\simeq 0.2$ mm, corresponding to an electron density of 3×10^{22} m^{-3}. This agrees well with the upper limit of 10^{22} m^{-3} on the electron density given by the linewidth measurements (although in this rough estimate recombination has not been considered, and the assumptions made could easily exhibit a large error).

For LTE to exist the transfer time, t, of energy between between electrons and ions must also be rapid compared to the duration of the laser pulse, and is given with good approximation (Ready 1971) by

$$t = 25T^{3/2}M/nZ^2$$

where T is the plasma temperature, M the relative atomic weight of the ions, Z the degree of ionization of the plasma and n the electron density. For $n \geq 10^{15}$ cm^{-3}, $t \leq 1$ ns. A LTE approximation is therefore appropriate, which makes a quantitative analysis possible.

In a second example, LIBS spectra were obtained for Brass in the range 400 nm to 525 nm, and a typical spectrum is shown in figure 4. A laser pulse energy of 20 mJ and a gate delay of 80 ns were used, and the target was stepped every 40 laser shots. Line intensities of several CuI transitions were measured, and the corresponding transition probabilities and degeneracies were obtained from the literature (Handbook of Chemistry and Physics 1984). Values for $\ln(I\lambda/Ag)$ were

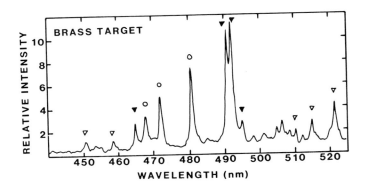

Figure 4. LIBS spectrum for Brass target, 80 ns after excitation; ▼ = CuII lines, ▽ = CuI lines, O = ZnI lines

then calculated and plotted against the upper state energy, shown in figure 3b. A plasma temperature of 14100 K (1.2 eV) was obtained from the gradient of the resulting curve. Again, a straight line fits the data well, suggesting that a single temperature can be used to characterize the conditions in the plasma, and allowing quantification of results.

Finally, spectra were also obtained from an aluminium target, in the range 300 nm to 710 nm. The laser pulse energy was 30 mJ, and the target was stepped after every 50 laser shots. The spectra obtained for 200 ns and 300 ns Boxcar gate delays are shown in figure 5, in which 19 peaks can be identified as resulting from AlI, AlII or AlIII transitions. It can be seen that, even at the relatively lengthy time delay of 200 ns, the noise level is high and the ionic lines are broadened significantly, making intensity measurements difficult. Nevertheless, AlII line intensities and reference A and g values were used to construct a Boltzmann plot, shown in figure 3c. It can be seen that a straight line does not fit the data well, suggesting that the plasma has expanded and possibly recombined to a degree where the dominant energy transfer mechanism is radiative decay, rather than collisions with electrons, rendering the LTE approximation invalid.

This was confirmed when a gate delay of 300 ns was used in the recording of the spectrum; a Boltzmann plot from the same lines again gave a poor fit, and furthermore suggested that the apparent plasma temperature had *increased* compared to the value calculated from the 200 ns delay results.

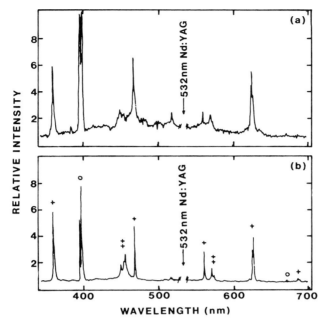

Figure 5. LIBS spectrum for aluminium target, at (a) 200 ns and (b) 300 ns after excitation; ++ = AlIII lines, + = AlII lines and o = AlI lines

3.2 Time Evolution of Spectral Features

In order that the optimum conditions for observation of the spectral emission can be determined, information concerning the temporal behaviour of the plasma is vital. The LTE assumption is likely to become less valid at long time delays, where the plasma has expanded to an extent where radiative decay becomes the dominant energy transfer mechanism; on the other hand the quality of the spectra, in terms of signal–to–noise ratios and broadening of spectral lines, improves as the plasma cools. Therefore, from an analytical viewpoint, the instant at which the emitted intensity peaks for a particular species may not be the optimum time for observation.

For the NaCl target spectra were taken at gate delays of 200 ns and 400 ns (see section 3.1), and from the corresponding Boltzmann plot it is deduced that the LTE approximation is valid at the shorter gate delay, where the spectrum still shows good signal–to–noise ratios. A delay of around 200 ns was therefore used in attempting to obtain quantitative results (see section 3.3 below).

LIBS Spectra for the Brass target were obtained in the wavelength range 420 nm to

525 nm, for a laser pulse energy of 20 mJ, with gate delays between zero and 100 ns in 20 ns steps, shown in figure 6. The broad band Bremsstrahlung radiation is apparent at short delays, with the subsequent evolution of neutral and ionic emission lines. The line doublet (attributed to CuII) at $\simeq$ 492 nm is not yet resolved as such in the 40 ns delay spectrum, due to broadening of the peaks at high plasma density, and the signal-to-noise ratio can be seen to improve as the delay between the laser pulse and observation gate increases. A gate delay of 80 ns was considered optimum, since at shorter delays the noise level prevented accurate measurement of line intensities. The intensity measurements were therefore taken from the 80 ns delay spectrum.

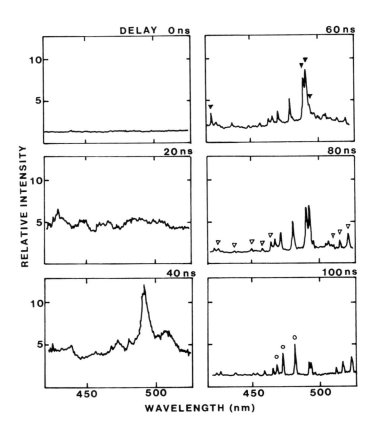

Figure 6. LIBS spectra at different times after excitation for a Brass target; ▼ = CuII lines, ▽ = CuI lines and O = ZnI lines

The results for an aluminium target demonstrated that conditions of LTE did not hold at delays of 200 ns or more; however for the purposes of spectral analysis it was not possible to reduce the delay below this, since the high background/noise levels and broadening of spectral features made measurement of line intensities impossible. Thus, in attempting to obtain any quantitative information from the Al spectra, it must be borne in mind that a LTE assumption may not be correct.

The presence of spectral features resulting from transitions in neutral, singly and doubly ionized aluminium did enable a direct comparison of the temporal characteristics of the emissions from the various species, by scanning the Boxcar gate delay at wavelengths of 394.40 nm (AlI), 466.3 nm (AlII) and 452.9 nm (AlIII). The delay range 0 – 1 μs was scanned with a gate width of 25 ns, taking 100 data points across the line profile. All curves were normalized to give a maximum relative intensity of 1.0, and are shown on the same graph in figure 7. The quoted transition probability for the 394.40 nm line is 0.5×10^8 s^{-1}, which corresponds to an excited state lifetime of 20 ns.

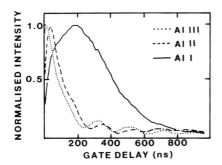

Figure 7. Time scans of AlIII, AlII and AlI peaks

It can be seen that the emitted intensity does not peak until 200 ns after the laser pulse. A possible explanation for this is that the excited state population is being augmented by decay from higher lying states, lengthening the observed lifetime; on the other hand the overall number of neutral AlI atoms will also increase with time due to recombination of ions and electrons.

A factor which will affect time resolved measurements is the response of the PMT/amplifier combination; in the traces for AlII and AlIII small oscillation 'peaks' occur at 175 ns intervals, being evidence of 'ringing' (this is hidden by the large signal at long delay in the AlI trace). The rise time of the amplifier is adequate, since the emission at 452.9 nm shows a peak in intensity only 25 ns after the laser pulse; the similar shape of the decay curves for the three signals indicates that there most likely is an instrumental effect occurring, rather than a true representation of the decay of the various emissions.

3.3   Elemental Analysis

We attempted to derive elemental constituents for all targets discussed thus far.

The ratio of Na to Cl from the NaCl target should yield a composition of 50:50. Ideally one would derive this from measuring the intensities of adjacent NaI and ClI lines; however, this proved not to be possible because of the lack of sufficiently strong ClI lines in the spectral range examined, and because of the spectral overlap with the wings of the much more prominent ClII and NaI lines. Furthermore, the noise level at relatively short delays made the measurement of weak lines difficult; increasing the gate delay would reduce this noise but would render the LTE approximation invalid.

For the Brass spectrum at a gate delay of 80 ns the ratio of the most abundant constituents, copper and zinc, was determined from the measured intensities of the CuI line at 465.11 nm and the ZnI line at 481.05 nm, using $g.f$ values obtained from the Handbook of Chemistry and Physics (1984) and Russell (1933), for a plasma temperature of 1.2 eV. A value of 66.0 % ($\pm$ 12 %) was obtained for the Cu concentration, which is in reasonable agreement with the accepted value of 75 % ($\pm$ 3 %) for 'yellow' Brass, considering the many possible sources of error, including inaccuracies in quoted oscillator strengths, errors in intensity measurements due to noise and background, the estimated plasma temperature, the necessary assumption that LTE holds (which affects the accuracy of the calculated Boltzmann distribution), and finally the error caused by variations in the ablation laser pulse energy. This last factor was significant in our LIBS system, since in addition to the inevitable shot to shot variation in the laser pulse energy our Nd:YAG system exhibited a noticeable periodic intensity fluctuation. In a truly analytical instrument a method of stabilizing the output intensity of the ablation laser would be necessary to avoid the complications of calibrating the signal against the laser power.

From the aluminium spectrum shown previously in figure 5 it can be seen that there are some peaks which belong to magnesium, an element common to aluminium alloys in fairly low concentrations (the peaks also occur in the spectrum shown in figure 8). An estimate of the Mg concentration was made from the spectrum by measuring the intensity ratio of two lines in relatively close proximity, the magnesium line at 383.23 nm and the aluminium line at 394.40 nm (a plasma temperature of the order of 1.5 eV (17000 K) is assumed). Using this value, together with the measured line intensities, excited state lifetimes and $g.f$ values obtained from Striganov (1968) and Griem (1964), a concentration of 0.63 % (± 0.3 %) of magnesium in the aluminium alloy was obtained (usual concentrations of magnesium in a typical aluminium alloy are in the range 0.1 % to 5 %). The accuracy of this value will be affected by the almost certainly inaccurate assumption that LTE holds, in addition to the factors mentioned above.

In order to investigate the possibility of detecting surface contaminants fine rhenium powder (particle diameter 3 μm) was deposited on the surface of the aluminium alloy disc. The density of Re on the surface was around 10 mg cm^{-2} (3 × 10^{19} atoms cm^{-2}), which corresponded to 6 μg of Re within the laser focal area. The laser pulse energy was around 15 mJ. A monochromator wavelength increment of 0.225 nm was used, since a large wavelength range between 300 nm and 700 nm was scanned. Each data point consisted of the averaged signal over 30 laser shots. A gate delay of 300 ns and a gate width of 25 ns were used to avoid recording the intense background radiation which occurred at shorter time delays. The spectrum obtained is shown in figure 8, together with a scan for a clean aluminium alloy target at the same gate delay to facilitate direct comparison.

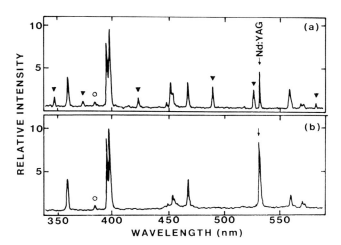

Figure 8. LIBS spectra, observed 300 ns after excitation, for (a) aluminium target with rhenium surface coating and (b) pure aluminium target. ▼ = rhenium lines, O = magnesium lines

A value of 9.6 % (± 3 %) was obtained for the rhenium/aluminium ratio, by measuring the intensities of the ReI line at 346.05 nm and the AlI line at 394.4 nm; oscillator strengths were obtained from Duquette (1982). It should be noted that, since the rhenium was only present on the surface of the aluminium target, this figure is only meaningful as an indication of the *average* Re/Al ratio in the laser produced plasma; we expect the first few shots to contribute greatly to the Re signal, while subsequent laser pulses on the same area of the target will produce the strong aluminium emission lines observed. The figure of around 10 % does, however, suggest that most of the rhenium powder is removed in the first 3 or 4 laser shots (30 shots in all were incident on each spot on the target), despite the average depth of powder on the surface of around 100 μm.

4. CONCLUSION

The technique of laser induced breakdown spectroscopy has several advantages for trace element detection at the percentage level. The need for little or no sample preparation and potential for simultaneous multi-element detection, coupled with the simplicity of the method and spatial resolution of the order of a few microns make LIBS an attractive possibility for rapid (on-line) analysis, although for this to be realised the use of a gated photodiode array would be vital. The vacuum environment chosen ensured that, at the levels of interest, spectral interference from buffer gases or atmospheric constituents was eliminated. We discuss this in greater depth (Allott *et al* 1990) in a report on work investigating the effects of the presence of buffer gases or atmospheric environments.

The principle limitations with the current arrangement are the scanning monochromator, which increases data acquisition times, and the poor stability of the ablation laser, which affects the accuracy of the measured line intensities. Beyond this the present investigation did highlight some difficulties which arise when trying to establish whether LTE exists. The results demonstrate that the assumption of LTE, necessary for any quantitative analysis, must be treated with some care due to the dependence on many factors, with the choice of delay between the ablation laser pulse and observation gate among the most important.

The experiments using an aluminium disc coated with rhenium powder indicate that analysis of surface contaminants is possible, and furthermore demonstrate that the LIBS method is not limited to solid samples but that a material in loose powder form can be investigated, without the need for adhesives or compression. The results also demonstrate that simultaneous multi-element analysis is easily achieved.

ACKNOWLEDGEMENT

Financial support for part of this work by AWE, Aldermaston, is gratefully acknowledged.

REFERENCES

Adrain R S and Watson J 1984 J. Phys. D **17** 1915

Allott R M, Goddard B J and Telle H H 1990 Spectrochim. Acta B (to be published)

Duquette D W, Salih S and Lawler J E 1982 J. Phys. B **15** L897

Griem H R 1964 Plasma Spectroscopy (New York:McGraw–Hill)

Handbook of Chemistry and Physics 1984 (Florida: CRC Press)

Krupp GmbH Essen 1990 Press release for Hannover Fair

Laqua K 1985 analytical Laser Spectroscopy eds S Martellucci and A N Chester (New York:Plenum) 159

Leis F, Sdorra W, Bak Ko J and Niemax K 1989 Mikrochim. Acta [Wien] **II** 185

Ready J F 1971 Effects of High–Power Laser Radiation (New York: Academic Press)

Russell H N 1933 Astrophys. J. **78** 239

Schueler B and Odom R W 1987 J. Appl. Phys. **61** 4652

Striganov A R and Sventitskii N S 1968 Tables of Spectral Lines of Neutral and Ionized Atoms (New York:IFI/Plenum)

Tambini A J and Telle H H 1990 Laser Chemistry (to be published)

Tozer B A 1976 Opt. and Laser Tech. **8** 57

Optogalvanic spectroscopy as a diagnostic of operating fluorescent lamp electrodes

M.Duncan and R.Devonshire, High Temperature Science Laboratories, Chemistry Department, Sheffield University, Sheffield S3 7HF, U.K.

ABSTRACT. This paper describes preliminary results from an investigation of the cathodes of operating fluorescent lamps using photoemission optogalvanic spectroscopy (POGS).

1. INTRODUCTION

The life of a fluorescent lamp is largely determined by the life of its electrodes. These electrodes consist of a coil of very fine tungsten wire (with a diameter of about 0.1 mm) which is wound on a mandrel to produce a secondary coil with an outer diameter of approximately 0.5 mm. This coil is impregnated with an emissive material, which is a mixture of several metal oxides, BaO, SrO, CaO and ZrO_2. The impregnated coil is then loosely coiled to produce the final electrode configuration which is typically 14 to 15 mm in length when mounted upon rigid wire supports. During the life of the lamp the electrodes deteriorate as emissive material is evaporated and sputtered into the plasma surrounding the electrodes. It is this depletion of emissive material which determines the life of the lamp. The removal of emissive material along an electrode is not uniform, nor is it simply related to the electrode's temperature profile. On first starting a lamp a hot spot forms on the negatively biased side of the electrode. The hot spot corresponds to the point where the arc "attaches" itself to the electrode. During the life of the lamp this hot spot migrates across the electrode as the emissive is removed, leaving bare tungsten electrode material behind it.

Only a relatively small number of laser diagnostic experiments have been applied to the study of the important electrode region of an operating fluorescent lamp. Bhattacharya (1989a, 1989b) used laser induced fluorescence (LIF) techniques to determine the number densities of neutral and ionised barium atoms in the plasma surrounding the cathode. Hilbig and Chittka (1989) used saturated laser absorption (SLA) to address the same problem. Of more direct relevance to the present study is the work of Schulman and Woodward (1989) who used optogalvanic spectroscopy (OGS) as the basis for an **in situ** electrode diagnostic technique for operating fluorescent lamps. Their experiments were designed to measure the local effective work functions of the operating cathode surface; the technique having previously been described by Downey et al (1989) and referred to, by him, as photoemission optogalvanic spectroscopy (POGS). POGS signals are induced when a laser beam is incident on the cathode surface in a discharge system; this causes photoemission of electrons which are amplified in the local electric field to create an optogalvanic signal.

This paper describes preliminary results from an investigation of the cathodes of operating fluorescent lamps using POGS and associated measurements of electrode temperature profiles.

2. EXPERIMENTAL

The experiments are being undertaken on T12 (20 W, 610 mm long and 51 mm in diameter) and T8 (18W, 610 mm long and 38 mm in diameter) fluorescent lamps from which the phosphor coating stage of production has been omitted to allow the laser beams ready access to the electrode regions. The electrodes in the lamps did not have a regular geometry but were regarded as adequate for the purposes of these initial experiments. The lamps are mounted vertically (i.e. coil axis horizontal) on a precision translation stage and operated DC at a current of 200 mA (I_L) using a stabilised power supply, (Farnell Model E.350 Mk-II) and an appropriate ballast. An additional isolated DC power supply (Kingshill Model 18V10C) is used for cathode heating (I_H) (see **Figure 1**).

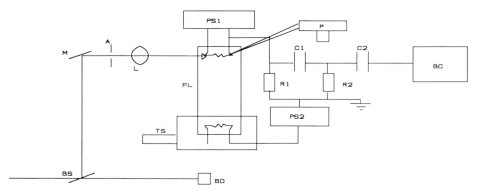

Figure 1. A schematic of the apparatus.

A Aperture, **BC** Boxcar, **BD** Beam Dump, **BS** Beam Sampler, **C** Capacitor, **FL** Fluorescent Lamp, **L** Lens, **M** Mirror, **P** Pyrometer, **PS** Power Supply, **R** Resistor, **TS** Translational Stage

The third harmonic (355 nm, 3.49 eV) from a Nd:YAG laser (Spectra-Physics single-mode Gaussian optic DCR-3) was sampled (using a simple glass plate) and further attenuated using neutral density filters to produce a POGS probe beam with a pulse energy of approximately 0.35 mJ (the laser system runs at 20 Hz with an output pulse with a temporal FWHM of 7 ns). This beam is focused a few millimetres before the electrode (see **Figure 2**). (The probe beam is used also to align the experiment itself since the transmitted beam, when projected onto a screen, gives a clear, magnified, image of the region of the electrode under investigation). Two focusing lenses have been used in the work. Initial experiments on T12 lamps employed a convex lens producing a spot incident on the cathode with a diameter of about 1 mm. When translating the lamp horizontally to investigate different regions of the electrode it was necessary to adjust the probe beam optics to ensure good overlap between the probe beam and the tertiary coils of the cathode. Subsequent experiments on T8 lamps, which have smaller 10 mm cathodes, measured along the direction of the electrode axis, have employed a cylindrical lens to generate a vertical region of illumination at the cathode, 6 mm in height and 0.5 mm wide. This gives improved spatial resolution at the electrode and eliminates the need to adjust the probe beam optics when translating the lamp to look at the POGS signals from different regions of the electrode.

The voltage across the lamp displays a saw-tooth waveform with an amplitude of approximately 2 V and a frequency of about 1 kHz. These are believed to be anode oscillations as described by Waymouth (1971). Initially, POGS signals were observed, and their magnitude recorded, using an oscilloscope. Subsequently the transient optogalvanic signals were recorded using an RC high pass filter circuit and a boxcar detection system (SRS, Model 250).

The temperature profiles along the length of the operating filaments were recorded using a photoelectric pyrometer (Minolta Land, Cyclops 52) with the emissivity set to the value appropriate for a bare tungsten surface. The pyrometer was fitted with a lens which gives a target diameter of approximately 2 mm. These measurements must be regarded as approximate for a number of reasons: calcium and barium atoms present in the negative glow of the discharge emit red radiation which can be detected by the pyrometer; the emissivity of tungsten surfaces impregnated with metal oxides, and in various stages of depletion, are unknown; and, finally, there are pyrometer reading correction factors which remain to be determined. Nevertheless, the uncorrected pyrometer measurements are adequate to display the trends in temperature values necessary to support the arguments presented in the present paper.

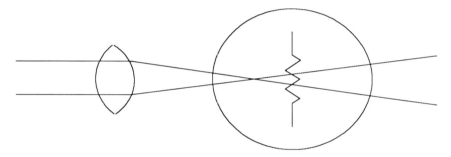

Figure 2. The relative positions of the probe laser focal region and the cathode.

3. RESULTS AND DISCUSSION

Figure 3 shows both the temperature, and the magnitude of the POGS signal, along the length of the cathode of an operating T12 lamp. The maximum in the temperature corresponds closely with the arc attachment point, which is the region of maximum current density in the plasma (and therefore of maximum surface electric field) and this is just the point where a maximum in the POGS signal is observed.

Increasing the external cathode heating current (I_H) much beyond 300 mA is found to significantly perturb the temperature profile of the cathode. It may be expected that as the average temperature of the cathode increases the thermionic emission from the cathode will also increase. The fixed discharge current will then be supplied at a lower surface electric field (or cathode fall). Thus an increase in I_H should decrease the cathode fall and a decrease in the magnitude of the POGS signal should be detected. This explanation is supported by the experimental results given in **Figure 4**. These show the variation in the temperature and the laser induced POGS signals from a point on the cathode near to the arc attachment point (which was about 3.5 mm from the negatively biased side of the cathode during these measurements) with increasing I_H.

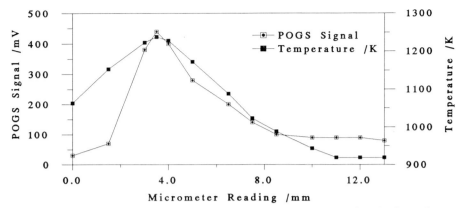

Figure 3. The temperature profile, and the laser induced POGS signal, along the length of the cathode axis in a T12 lamp; cathode heating is 200 mA.

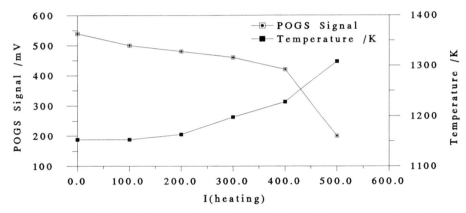

Figure 4. Variation in the temperature and the laser induced POGS signals at a point near the arc attachment point (about 3.5 mm from the negatively biased side of the cathode) with increasing heating current (I_H).

When I_H is increased to a point where the fixed discharge current (200 mA) is supplied through thermionic emission from the cathode then the surface electric field is effectively zero. This condition makes it possible to estimate the average zero field thermionic emission (ZFTE) work function of the cathode. On the basis of the I_H value, from **Figure 4**, where the POGS signal reaches zero, an estimate of 2 eV is obtained, via the Richardson equation, for the ZFTE work function of the electrode at this point. This value is in general agreement with the work of Waymouth (1971) where it was estimated that the average ZFTE work function for an operating fluorescent lamp cathode is in the range 1-2 eV.

Figure 5 shows three POGS profiles for the 10 mm long cathode in an operating T8 lamp, recorded successively over a period of approximately 1 hour. The results demonstrate very well the general level of reproducibility of the POGS measurements performed by us. The profiles shown in **Figure 5** were recorded in experiments where the cylindrical lens was used. Each point in the figure represents an average over 30 laser shots. The background noise level in the experiment is about 20 mV and arises primarily from electrical pick-up originating from the laser's Q-switch.

As mentioned in the introduction of this paper the life of a fluorescent lamp is largely determined by the life of the electrodes. It is known that significant damage is experienced by the electrodes each time the lamp is switched on, the damage arising from the physical loss of cathode material due to sputtering. **Figure 6** compares the stabilised POGS profiles of the cathode of a T8 lamp (not the same lamp as used for the data in **Figure 5**). The "before" profile was obtained after the lamp had been running for about 15 minutes and the "after" profile when it had stabilised for a couple of minutes after a lamp off-on on cycle. Given the reproducibility of POGS signals for stable running conditions demonstrated by the results in **Figure 5**, the profiles are seen to be significantly different. The result is a good indication of the utility of POGS measurements in cathode diagnostics.

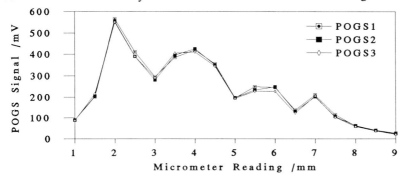

Figure 5. Three successive POGS profiles for the 10 mm long cathode of a T8 lamp.

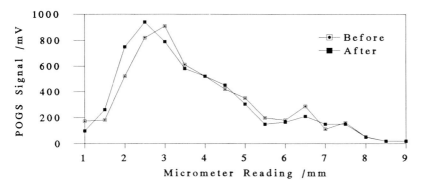

Figure 6. Stabilised POGS profiles of the cathode of a T8 lamp, (not the same lamp as used for data in **Figure 5**). The "before" profile was obtained after the lamp had been running for about 15 minutes, and the "after" when it had stabilised for a few minutes following a single lamp off-lamp on cycle.

To investigate, even more directly, the effect of sputtering on the POGS profiles, the second harmonic output (532 nm, 2.33 eV) from the Nd:YAG laser was used to ablate the emissive surface coating from a central region of the cathode. A 532 nm beam with a pulse energy of 11.5 mJ was focused to a point a few millimetres before the cathode, such that the diverging beam was incident on the cathode with a spot size of approximately 2 mm. Both the 355 nm probe beam and the 532 nm ablation beam were aligned parallel to each other, to give the 532 nm beam access to the cathode without modifying either the lamp's position or the probe beam's optics. Firstly, the POGS profile was recorded before the ablation laser beam was allowed to strike the cathode. The ablation beam was then allowed to strike the cathode for 30 s. (A POGS signal generated during this period was approximately four times greater than the maximum signal generated with the usual 355 nm probe beam). The profile was recorded again after a few minutes delay to allow the POGS signal to settle down. The two contrasting profiles obtained in this way are shown in **Figure 7**. Given the reproducibility of the POGS signals, demonstrated in **Figure 5**, there is seen to have been a dramatic change in the POGS profile induced by the ablation of cathode material. The result again illustrates the potential of POGS measurements for electrode surface diagnostics.

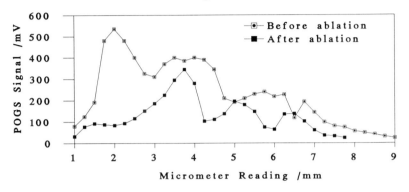

Figure 7. POGS profiles before and after laser ablation of the surface coating from the central region of the coating of a T8 lamp.

SUMMARY

This paper describes preliminary results from an investigation of the cathodes of operating fluorescent lamps using photoemission optogalvanic spectroscopy (POGS). The results have demonstrated clearly the high degree of reproducibility of POGS measurements on stable operating lamps and their sensitivity to changes in electrode characteristics. POGS measurements are believed to have potential in diagnostics of the surface condition of electrodes in operating fluorescent lamps. A complete account of this work is to be published (Duncan and Devonshire 1991).

ACKNOWLEDGEMENTS

The authors gratefully acknowledge useful discussions with D O Wharmby throughout the course of this work and THORN Lighting Ltd. (Light Sources Division) for providing the lamps. MD acknowledges financial support from the S.E.R.C. and THORN Lighting Ltd.

REFERENCES

Bhattacharya A K 1989a J. Appl. Phys. **65**(12) 4595
Bhattacharya A K 1989b J. Appl. Phys. **65**(12) 4603
Downey S W, Mitchell A and Gottscho R A, 1988 J. Appl. Phys. **63**(11) 5280
Duncan and Devonshire (to be published)
Hilbig R and Chittka U 1989 The Fifth International Symposium on the Science and Technology of Light Soures, Ed. by R Devonshire, J Meads and D O Wharmby ISBN 0-85426-034-X
Schulman M B and Woodward D R 1989 Appl. Phys. Lett. **55**(16) 1618
Waymouth J F 1971 Electrical Discharge Lamps, MIT Press, ISBN 0-262-23048-8

Optogalvanic spectroscopy of plasma processes and autoionization levels

R. Shuker and M. Hakham-Itzhaq.

Department of physics, Ben-Gurion University of the Negev, Beer-Sheva, Israel

ABSTRACT

The pulsed optogalvanic effect (OGE) and its utilization in spectroscopy are described in two different areas of research, exhibiting its power in atomic and discharge studies as examples for its uses. One of these studies involves the role of electron collisions versus heavy particle effects such as Penning ionisation in Hg-Ne hollow cathode discharges as a function of the Hg atom density. As the current increases the vapour pressure of Hg increases resulting in the domination of the Hg, rather than Ne, in determining the discharge characteristics. We describe the use of the time resolved pulsed optogalvanic signal (OGS) to determine the region in which a transition from Penning ionisation of Hg by Ne metastables to electron multistep ionisation of Hg occurs. The other example is the use of the time dependent shape of the OGS as a signature of the autoionising state. We also demonstrate the use of the OGS in studying linewidths and lineshapes involving such states.

1   Introduction

The optogalvanic effect (OGE) has long been used for atomic and molecular spectroscopy *(Pfaff et al 1984,Goldsmith and Lawler 1981 , Shuker et al 1983)*, and the study of plasma diagnostics and plasma processes in discharges *(Ausschnitt et al 1978)*. We have suggested a phenomenological model for the OGE *(Ben Amar et al 1983)* and studied such plasma processes as Penning ionization and others.

In the following section we detail a study of plasma processes in Hg hollow cathode discharges. Electron multistep impact ionization (EMSII) and the Penning ionization of the cathode metal vapour atoms are the main processes responsible for the discharge in hollow cathode discharge (HCD) tubes. At low current, the metal vapour atom density is relatively low, the high density of the inert gas, usually neon, controls the discharge and metal atom ionization is acheived through the Penning process. The EMSII process has a high rate only when the metal vapour density is relatively high, a condition which is obtained at high discharge current. The Hg/Ne HCD tube has exceptional behaviour. Here, a high density of mercury vapour occurs even at relatively low currents and a high EMSII rate exists and dominates the discharge. The transition between Penning and EMSII dominated ionization conditions is detected by the pulsed optogalvanic technique. We display optogalvanic signals which exhibit this transition behaviour.

In section 3, we demonstrate the use of the OGE in autoionisation spectroscopy. We have measured the line profile of two Copper quartet autoionisation levels. The optogalvanic signals which are detected from Cu $5s'\ ^4D_{5/2}$ and $5s'\ ^4D_{3/2}$ autoionisation levels are usually weak due to the low vapour density (10^{11} cm^{-3}) of the Cu vapour atoms inside the hollow cathode discharge tube operating in the steady state condition. Earlier measurements on autoionisation line intensities of copper were made by *Allen (1932)* and were carried out again forty years later by *Assous (1973)*. These measrements predict that the autoionisation transition probabilities are of the order of 10^{12} sec^{-1}. The linewidths associated with the $5s'\ ^4D_{5/2}$ and $5s'\ ^4D_{3/2}$ levels of Cu were calculated and are of the order of 6.5 cm^{-1} and 2.5 cm^{-1} respectively. In this work we have found that the lineshapes of the Cu autoionisation transitions have characteristic asymmetric Fano-shape line profiles. Both of these studies clearly emphasize the usefulness of the optogalvanic technique.

2. Processes in the Hg/Ne discharge

The pulsed resonant optogalvanic technique is used to study the discharge characteristics of the Hg/Ne hollow cathode discharge tube. In various HCD tubes such as Ca/Ne, Cu/Ne etc, the atomic vapour is produced by the sputtering process, usually in low densities on the order of 10^{13} cm^{-3}. As a result, the electron multistep ionization of the metal vapour atoms has a very low rate and its contribution to ionization and the production of free electrons is minor compared with other processes such as the Penning process. However, a Hg/Ne HCD tube provides a case in which a high density of vapour atoms is produced by thermal evapouration of the cathode at quite low currents. At high currents this density can exceed that of neon which is on the order of 10^{17} cm^{-3}. Subsequently the EMSII rate of the metal vapour atoms becomes quite high and dominates the discharge through increasing the total ionization rate above the ionization rate created by the Penning process. This enhancement is detected using the pulsed resonant optogalvanic technique.

2.1 Theoretical considerations

The discharge inside a hollow cathode tube involves the following main ionization processes of the metal vapour:

A) The Penning ionization process dominates the discharge at low currents when the metal vapour density is low. This interaction involves an energy transfer which occurs when an excited metastable atom of neon ionizes a ground state metal vapour atom . In the case of mercury it is according to the reaction

$$\text{Ne}^*(^3P_2, ^3P_0) + \text{Hg}(^1s_0) \rightarrow \text{Ne}(^1s_0) + [\text{Hg}^+(^2D, ^2P)]^* + e^- \ .$$

This is a quasi-resonant process. The levels of the neon and the mercury involved in the interaction are displayed in Fig 1. The rate of the Penning ionization is proportional to the transition rate given by the Fermi golden rule -

$$W(r) = 2\pi \ | < \text{Ne}, \text{Hg}^+, e^- \ | V(r) | \ \text{Ne}^*, \text{Hg} > |^2 \ \rho_f(\epsilon) / h \quad (1)$$

where V(r) is the interaction potential, r is the internuclear distance and $\rho_f(\epsilon)$ represents the final electron density of states. The whole process exhibits a quasi-resonant behaviour similar to other cases *Kane (1983), Kieffer et al (1966)*. The Penning ionization rate can be approximated by:

$$dn^P_i(t) / dt = \gamma_p \ n(\text{Ne}^*) \ n(\text{Hg}). \quad (2)$$

where, $n^P_i(t)$ is the number density of Hg ions due to Penning ionization, and $\gamma_p = <\sigma_p v_p>$ is the Penning ionization rate coefficient. According to *Wren et al (1981)* the Penning ionization cross-section $\sigma_p =$

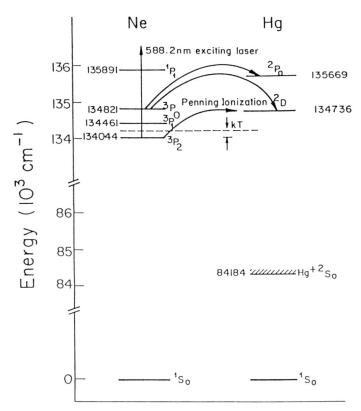

Figure 1 Neon and Mercury levels involved in the Penning ionization process.

$7.10^{15} cm^2$. We estimate $v_p = 4.5 \cdot 10^4$ cm/sec, $n(Hg) = 10^{14} - 10^{17}$ cm^{-3} and $n(Ne^*) = 5.10^{11}$ cm^{-3} *(Kane 1983)*.

B) The EMSII process involves ionization of the metal vapour atoms by electrons. This process has a high rate at high currents when the metal vapour density is high as is the case for mercury. The rate is given by the following equation

$$d\, n^D_i (t) / dt = K_e\, n_e\, n(Hg). \tag{3}$$

where, $n^D_i (t)$ is the number density of Hg ions due to EMSII and $K_e = <\sigma_e\, v_e>$ is the EMSII rate coefficient. For mercury atoms at $kT_e = 0.8$ ev, $\sigma_e = 10^{16}$ cm^2 *(Kieffer et al 1966)*, $n_e \cong 10^{10} - 10^{12}$ cm^{-3}, $v_e \cong 6.5 \cdot 10^7$ cm/sec and the EMSII rate coefficient is $K_e = 6.5 \cdot 10^{-9}$ cm^3/sec.

The rate term $d\,n^D_i(t)/dt$ at high currents is $10^{20}\text{cm}^{-3}\text{s}^{-1}$ and at low currents is of the order of $10^{17}\,\text{cm}^{-3}\text{s}^{-1}$. The EMSII rate is related to the electron energy distribution $n_e(\varepsilon)$ which is assumed for simplicity to be Maxwellian-like :

$$n_e(\varepsilon)\,d\varepsilon = [(2n_e/\pi^{1/2})]\,((\varepsilon/T_e)^{1/2})\,\exp\{-\varepsilon/kT_e\}\,d(\varepsilon/kT_e). \quad (4)$$

n_e is the total electron density ($n_e \approx 4 \cdot 10^{11}\,\text{cm}^{-3}$) and $kT_e \approx 0.8$ ev. In the following we discuss briefly the optogalvanic technique *Ben Amar et al 1981, 1983 and Shuker et al (1983)*. The optogalvanic voltage signal (OGS) is given by :

$$\Delta V(t) = -\beta \Sigma_i\, a_i\, \Delta n_{(i)}(t) \quad (5)$$

β is a sensitivity factor which depends on the factor K_e and is given by $\beta = -\partial K_e/\partial V$ and a_i are the changes of the multiplication factor as a result of changes in the temporal populations of the excited states $\Delta n_i(t)$. These a_i coefficients are related to the ionization rate coefficients of the levels i and in general this is larger whenever the atoms are in excited levels close to their ionization limits. The rate equations for the population density deviations from steady state conditions, $\Delta n_i(t)$, in the relevant four-excited-states-theory *Ben Amar et al (1981 1983) and Shuker et al (1983)* are given by :

$$d(\Delta n_i)/dt = -\Delta n_i/T_i + \Sigma_{i\neq j}\,[\gamma_{ij}\,\Delta n_j - \gamma_{ji}\Delta n_i]. \quad (6)$$

where γ_{ji} are the relevant elements of the rate matrix and n_i are the excited state population densities that are involved in the optogalvanic effect. The population densities of these levels in steady state conditions are designated by n^0_i ; the deviation from the steady state is Δn_i and the densities n_i are given by $n_i = n^0_i + \Delta n_i$. We assume that the population n_i relaxes to n^0_i with a relaxation time T_i. These relaxation times are indicative of the coupling strength and the response of these levels to changes in the plasma as a whole. The four levels of our Hg/He are shown in Fig 2. These levels are : the neon metastable level $^3P_2$ and the excited level $^1P_1$, the ground state $^1S_0$ of Hg and the Hg$^+$ ion levels ^2P or ^2D. Because of the high density of Hg atoms at high temperatures we modify the resonant pulsed optogalvanic phenomenological model previously discussed

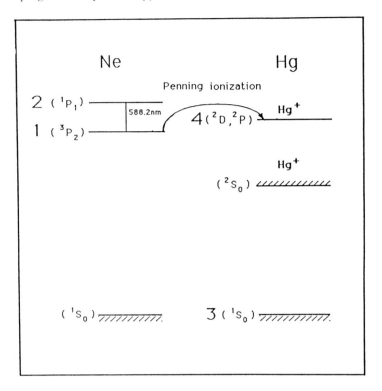

Figure 2 The four energy levels of the Ne/Hg system that give major contributions to the optogalvanic effect in the presence of Penning ionization of Hg by excited metastable neon atoms.

by *Ben Amar et al (1981, 1983)* and *Shuker et al (1983)*, by taking into consideration the case in which the high density of mercury vapour increases the rate of the EMSII process so that it dominates the discharge. We include the EMSII process in our model for the OGS by introducing a term $K_e n_e \Delta n_3$ which is the effective EMSII rate term. According to equation (6) the rate equations for the four relevant levels are:

$$d (\Delta n_1)/dt = - [1/T_1 + \gamma_p n_3] \Delta n_1 \qquad (7)$$

$$d (\Delta n_2)/dt = - [1/T_2] \Delta n_2. \qquad (8)$$

$$d (\Delta n_3)/dt = - \gamma_p n_3 \Delta n_1 - [1/T_3 + K_e n_e + \gamma_p n_1] \Delta n_3. \qquad (9)$$

$$d (\Delta n_4)/dt = \gamma_p n_3 \Delta n_1 + [K_e n_e + \gamma_p n_1] \Delta n_3 - [1/T_4] \Delta n_4. \qquad (10)$$

The initial values, after a laser pulse at resonance with the transition between levels 1 and 2 and which is much shorter than any time scale of the discharge, are :

$$-\Delta n_1(0) = \Delta n_0 = \Delta n_2(0) = Q_{12}(n_1 - n_2) > 0. \quad (11a)$$
$$\Delta n_3(0) = \Delta n_4(0) = 0. \quad (11b)$$

where the term Q_{12} describes the total absorption of a short laser pulse resonant with the transition between the levels 1 and 2 and is given by

$$Q_{12} = \int \sigma_{12} I(t) dt \quad (12)$$

where σ_{12} is the absorption cross section and $I(t)$ is the laser intensity. At high currents the rate term $\gamma_p n_1 \Delta n_3 < K_{en} e \Delta n_3$ and its contribution to the rate equations can be neglected.

The solutions of the rate equations are :

$$\Delta n_1(t) = -\Delta n_0 \, e^{-t/T_1^*} < 0. \quad (13)$$
$$\Delta n_2(t) = \Delta n_0 \, e^{-t/T_2} > 0. \quad (14)$$
$$\Delta n_3(t) = \{\gamma_p n_3 \Delta n_0 / \alpha_2\} [\, e^{-t/T_1^*} - e^{-t/T_3^*} \,] \geq 0. \quad (15)$$
$$\Delta n_4(t) = \{\gamma_p n_3 \Delta n_0 / \alpha_2\} \{[K_{en_e}((\, e^{-t/T_1^*}\,) / \alpha_1 - (\, e^{-t/T_3^*} / \alpha_3\,)) - \alpha_2 e^{-t/T_1^*}/\alpha_1 \,] - [(K_{en_e} - \alpha_2)/\alpha_1 - K_{en_e}/\alpha_3] * e^{t/T_4} \quad (16)$$

T_1^* is an effective plasma relaxation time for the neon metastable level which includes the Penning process, assuming n_3 is a slowly varying quantity, and where $1/T_1^* = 1/T_1 + \gamma_p n_3$. T_2 is an effective plasma relaxation time for the neon $^1P_1$ excited level while T_3^* is related to plasma relaxation time T_3 and the EMSII rate and is given by the term $1/T_3^* = 1/T_3 + K_{en\ e}$. T_4 is a diffusion relaxation time and represents the heavy particle ionic diffusion time and would be the smallest relaxation term.

The α_i terms are defined as:

$$\alpha_1 = [1/T_4 - 1/T_1^*\}. \quad (17a)$$

$$\alpha_2 = [1/T_3^* - 1/T_1^*] = [1/T_3 + K_{en_e} - 1/T_1 - \gamma_p n_3] \quad (17b)$$

$$\alpha_3 = [1/T_4 - 1/T_3^*]. \tag{17c}$$

α_2 is positive only at very high currents. At long times, T_1 and T_2 are the short relaxation times and $\Delta n_1, \Delta n_2 \to 0$. The most significant contribution to the OGS comes from the terms Δn_3 and Δn_4 at longer times. Thus, at these times equation (5) renders

$$\Delta V = - \beta \{ a_3 \, \Delta n_3 (t) + a_4 \, \Delta n_4 (t) \}. \tag{18}$$

The coefficients a_3 and a_4 are related to the ionization rates of neutral atoms and ions of mercury. We assume that $a_3 > a_4$ as the ionization of neutral atoms of mercury requires lower energy electrons which exists in relatively high densities in comparison with the second ionization process which requires high energy electrons that exist in relatively low densities according to the electron energy distribution in the steady-state discharge.
The term $\Delta n_3(t)$ is always positive as can be seen from Eqs 15 and 17. It contributes the negative part of the OGS.

The relaxation time T_4, which is the diffusion characteristic, is longer than other relaxation times, thus $T_4 > T_3^*, T_2, T_1^*$ rendering $\alpha_3 < 0$ and $\Delta n_4(t)$ can be written as :

$$\Delta n_4(t) = \{\gamma_p \, n_3 \, \Delta n_0 / \alpha_1\} \{ [(K_e \, n_e)/(K_e \, n_e + 1/T_3 - 1/T_4)] - 1 \} [e^{-t/T_4}] \tag{19}$$

We see that $\Delta n_4(t) \leq 0$ as $1/T_3 > 1/T_4$ and contributes the positive part of the OGS. Both equation (15) and equation (16) contain the same multiplication term $\gamma_p \, n_3 \, \Delta n_0$ and simplifications are introduced by taking the reduced versions of (15) and (16).

$$\Delta n_3^*(t) = \Delta n_3(t) / \gamma_p \, n_3 \, \Delta n_0 . \tag{20}$$
$$\Delta n_4^*(t) = \Delta n_4(t) / \gamma_p \, n_3 \, \Delta n_0 . \tag{21}$$

We note $T_4 > T_3 \gg T_1$ therefore $1/T_1 \gg 1/T_3 > 1/T_4$ and we can drop the term $(-1/T_4)$ from the denominator of (19). Hence we obtain:

$$\Delta n_3^*(t) \cong [e^{-t/T^*_1} - e^{-t/T^*_3}] / [(1/T_3 + K_e \, n_e) - (1/T_1 + \gamma_p \, n_3)] \tag{22}$$
$$\Delta n_4^*(t) \cong T_1^* \{ [(K_e \, n_e / (K_e \, n_e + 1/T_3)] - 1 \} e^{-t/T_4} \tag{23}$$

At high cathode temperature the mercury vapour pressure is high and n_3 increases exponentially and as $\gamma_p n_3 > K_e n_e$, Eq. (20) can be reduced to :

$$\Delta n_3^*(t) \cong [e^{-t/T_3}] / (1/T_1^*) > 0. \tag{24}$$

Now we can compare the contributions of $\Delta n_3^*(t)$ and $\Delta n_4^*(t)$ that dictate the shape and the value of the OGS under the condition that $a_3 > a_4$ as discussed earlier.

2.2 Experimental setup

The experimental setup is displayed schematically in Fig. 3. It includes a

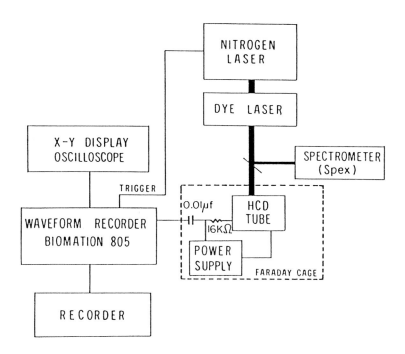

Figure 3 Experimental set-up used for the optogalvanic measurements

nitrogen laser with average power about 30 kW and pulse duration of 5nS. It pumps a nonflowing tunable Hansch-type dye laser. The dye laser has a pulse energy of a few microjoules, pulse duration of about 5 ns and a linewidth of 0.1 cm^{-1}. The setup consists also of an old Hg/Ne hollow cathode discharge tube (Pye Unicam type) filled with about 5 torr of neon driven by a very stable D.C. power supply to allow measurements of weak optogalvanic voltage signals (OGS) of only a few millivolts. To avoid R.F. interference, the discharge tube, its electrical circuit and the power supply were inserted in a Faraday cage. The OGS were coupled via a D.C blocking capacitor to a signal averager (Biomation 805) and displayed on an oscilloscope or on a chart recorder. The dye laser emission wavelength was monitored by a Spex 1401 double monochromator.

2.3 Experimental results and discussion.

The optogalvanic signal at the higher current of 1.3mA (Fig. 4), shows that

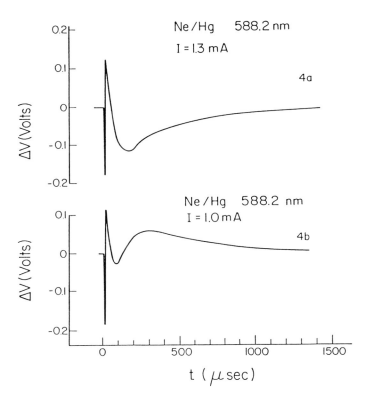

Figure 4 Optogalvanic signals taken from Ne/Hg HCD tube at various currents

there exists a region where the voltage across the discharge tube drops at a time longer than 150 μs after the laser pulse ends and reaches its lowest value 20 μs later. This indicates that the discharge current density increased although we have decreased the Penning ionization rate through laser illumination at 588.2 nm which depleted the metastable level. We suggest that the total ionization enhancement is created by an increase of the EMSII rate due to the additional atomic vapour of mercury that would have otherwise participated in the Penning ionization process. The major reasons that support our assumption that this domination occurs are :

a) The high vapour pressure of mercury dictates a lower electron temperature when the vapour density is higher than that of neon. This can occur when the cathode temperature is only about 480 °K (see Fig. 5)

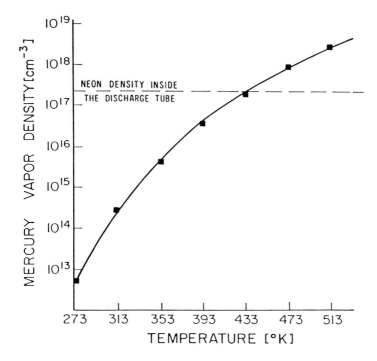

Figure 5 Vapour density of Hg calculated from CRC data handbook data (1971)

b) The multistep excitation of mercury through the electron impact ionization process starts at an electron excitation energy of 4.68 ev in comparison with an excitation energy of 16.67 ev required for the first excited state. According to equation (4) the density of such energetic electrons is very low.

c) A major contribution comes from the Hg singlet level (6p $^1P_1$) which is 6.7 ev above the ground state of Hg. Decay at this level is radiatively trapped and therefore it is heavily populated.

d) The first excited triplet of mercury contains two metastable states ($^3P_2$ and $^3P_0$) with long life-times which support the multistep ionization process. We modified the four-states model in order to take into account the increasing role of the mercury vapour atoms in dominating the discharge.

In the following we discuss various conditions of the discharge

A. <u>Currents (2 > i >1.2)mA</u>. This case has high cathode temperature and a high density of evapourated mercury exists. The EMSII process is the dominant ionization process in the discharge tube as $K_e n_e \gg 1/T_3$.

The term $\{[(K_e n_e) / (K_e n_e + 1/T_3)] -1\}$ in Eqn (23) is almost zero as $K_e n_e > 1/T_3 > 1/T_4$. We have $\Delta n_4(t) \to 0$ and $\{a_4 \Delta n_4(t)\} \cong 0$. The remaining term is $\{a_3 \Delta n_3(t)\}$, thus, the OGS shows a negative component; $\Delta V = - \beta\ a_3 \Delta n_3(t) < 0$.

This behaviour was observed experimentally and is demonstrated in Fig 4A. The measured values of T_i are : $T_4 = 500$ µs, $T_3 = 330$ µs, $T_3^* = 111$ µs, $T_1 = 10$ µs, $T_1^* = 0.34$µs and $T_2 = 7.2$µs. The value of T_1 is taken from the publication by *(Shuker et al 1983)*. It is estimated from the OGS in Na/Ne at i = 2 mA, where the Penning ionization process has not been observed. T_3 is calculated by substituting the estimated value of $K_e n_e \cong 6.10^3$ s^{-1} in the expression for T_3^*, namely, $1/T_3^* = 1/T_3 + K_e n_e$. We compare these data with the calcium data in which the relaxation frequencies *(Shuker et al 1983)* are as follows : $1/T_1^* = 1.5.10^5$ s^{-1}, $1/T_3^* = 1.9.10^4$ s^{-1}, $1/T_4 = 1.5.10^5$ s^{-1}, $\gamma_p n_3 = 1.10^5$ s^{-1}. We estimate $K_e n_e = 4 *10^3$ s^{-1} for calcium. Putting these values in equations (15) and (16) we obtain Δn_3 and Δn_4 for calcium. The ratio Δn_3(Ca) / Δn_3(Hg) is 0.26 and indicates that the calcium contribution to the negative

OGS is small compared with that of mercury. The recorded Ca optogalvanic signal *(Shuker et al 1983)* at 8 mA is in accord with our assumptions that the Ca density ($\sim 10^{14}$ cm^{-3}) is small compared with that of the mercury ($\sim 10^{17}$ cm^{-3}) thus the EMSII rate in calcium is low.

B. <u>Currents above 2mA</u>. Here the cathode is very hot and n_3 is high (C R C 52^{nd}1971) with respect to the neon density as seen from Fig 5. The term $\Delta n_3^*(t) = \Delta n_3(t) / \gamma_p n_3 \Delta n_0$ is very small, $1/T_3 > 1/T_4$ and $K_e n_e > 1/T_3$, and the term $\Delta n_4^*(t) \to 0$. The only contribution to the OGS that needs to be considered comes from $\Delta n_3(t)$.

$$\Delta n_3(t) = \{\gamma_p n_3 \Delta n_0 / \alpha_2\} [e^{-t/T_1^*} - e^{-t/T_3^*}] \geq 0, \quad (25)$$

since $\alpha_2 = [1/T_3 + K_e n_e] - [1/T_1 + \gamma_p n_3] \cong -[\gamma_p n_3]$. Thus the term $\Delta n_3(t)$ is small, as both the current and the time relaxation terms increase. The OGS at $t > 50\mu s$ is negative and small.

C. <u>Low currents (i < 1.2mA)</u>. Here the cathode temperature is low. Accordingly the evaporation rate of mercury $n_3 \cong 2.10^{14}$ cm^{-3} is low and the sputtered density of mercury (10^{14} cm^{-3}) is non-negligible. Here the Penning ionization process dominates the discharge. The justification for our last assumption is based on the estimation that $\gamma_p n_1 \Delta n_3 > K_e n_e \Delta n_3(t)$ as $n_e < 4.10^{10}$ cm^{-3}. The major contribution to the OGS comes from:

$$\Delta n_3^*(t) \cong [e^{-t/T_3^*}] / [\gamma_p n_3 - K_e n_e + 1/T_1 - 1/T_3]. \quad (26)$$

$$\Delta n_4^*(t) \cong T_1^* K_e n_e T_3^* e^{-t/T_4} - \{T_1^* e^{-t/T_4} + [K_e n_e T_1^* T_3^* e^{-t/T_3^*}]\}. \quad (27)$$

Experimentally, the detected OGS at $t < 100$ μs exhibits a moderate negative voltage signal and a huge positive voltage signal at $t > 100$ μs as is shown in Fig 4b. This behaviour of the OGS can be described mainly by Equations (26) and (27). The negative contributions that come from Equation (26) and the first part of Equation (27) at $t < T_4$ is partially diminished by the last term in Eq. (27). At $t > T_4$ the most significant contribution to the OGS comes from the second term of $\Delta n_4^*(t)$ and it is positive for additional positive contribution to the OGS results from the sputtering with a rate term I $\gamma_s \Delta n_4(t)$ *(Shuker et al 1983).*, which we have neglected earlier in the rate equations. Here, I is the discharge current and γ_S is the self-sputtering coefficient. Thus at low currents the OGS mainly exhibits a positive voltage signal that decays slowly as seen in Fig 4b. This is quite similar to the OGS detected in Ca/Ne tube *(Shuker et al 1983)*.

2.4 Conclusions.

The experimental OGS exhibits two different modes of the hollow-cathode discharge which are related to the current. These are related to Penning ionization and to EMSII. At currents below 1.2mA the OGS shows a Penning ionization signature which is similar to the case of Ca discussed by *Ben Amar et al (1983)*. At currents above 1.2mA the OGS exhibits a current density enhancement due to the high Hg atomic density resulting from thermal evaporation of Hg. In this case the EMSII process controls the discharge. We have expanded the resonant optogalvanic theory *Erez (1979), Ben Amar et al (1981,1983),* and *Shuker et al (1983)* to deal with a case in which the sputtering process has lost its major role of supplying the metal atomic vapour for the ionization processes to maintain the discharge in the steady state. This is the state where the EMSII process increases the current density and controls the discharge.

3. AUTOIONISATION SPECTROSCOPY

Autoionisation levels are generally very broad, short lived and lead to characteristic asymmetric line profiles, of the well known Fano-shape. We present here typical measurements of the lineshapes of autoionisating transitions in the of Cu atom using the pulsed optogalvanic technique, whereby the discharge response to changes in the populations and electron distributions induced by laser absorption is ultilized as a spectroscopic tool. We used a 15 nsec-7 Ghz linewidth pulsed dye laser to scan the transition profile of two autoionisation transitions in copper. A one cm^{-1} tuning of the dye laser is used across the broad transition profiles. The results exhibit the characteristic Fano shape for these transitions. This demonstrates the usfulness of this technique in the study of lineshapes and line-profiles.

3.1 The experimental setup.

The experimental setup is similar to that displayed in Fig 3 with some changes. It includes an excimer laser with average power of about 5MW and pulse duration of 15nS. It pumps a nonflowing tunable Hansch-type dye laser. The laser has a pulse energy of a few milijoules , pulse duration of about 15 nsec and a linewidth of 0.23 cm $^{-1}$. The setup also consists of

a Cu/Ne hollow cathode discharge tube (Hamamatsu type) filled with about 5 torr of neon driven by a very stable D.C. power supply to allow measurements of weak optogalvanic voltage signals (OGS) of only a few millivolts. The electrical circuit and the power supply are inserted inside a Faraday cage in order to avoid R.F. noise coupled to the optogalvanic signals. The OGS were coupled via a D.C blocking capacitor to a digital oscilloscope (HP 54020A) and displayed on an ink jet printer The dye laser emission wavelength was monitoried by a Spex 1401 double monochromator and a sensitive photodiode .

3.2 Theoretical considerations.

The configuration $3d^9 4s\,(^3D)\,5s$ of the copper atom contains four levels $5s'\,^4D_{1/2}$, $5s'\,^4D_{3/2}$, $5s'\,^4D_{5/2}$, $5s'\,^4D_{7/2}$. The energy interval ratios are 7 : 5 : 4. In the pure LS coupling scheme the interval energy ratios would be 7 : 5 : 3. Therefore intermediate coupling is more adequately used to describe the interval energy ratios of the highly excited levels of the copper atom. The copper autoionisation levels $5s'\,^4D_{5/2}$ and $5s'\,^4D_{3/2}$ are 632 cm^{-1} and 1267 cm^{-1} above the ionization level of the copper atom which is 62317 cm^{-1} above the ground state of Cu *(Shenstone 1948)* . Figure 6 shows the relevant energy levels and transitions.

The autoionisation spectral lines are very broad due to configuration interaction in which the discrete atomic level is coupled to a continuum state belonging to a different configuration. The theoretical interpretation is due to *Fano (1961)* who calculated the spectral distribution of the transition probability (or absorption strength) to a discrete state perturbed by the continuum. The results can be expressed in terms of two parameters, the value of the width Γ of the autoionization state and $(2\pi\Gamma/h)$, the autoionization probability. The second parameter is the Fano asymmetrical

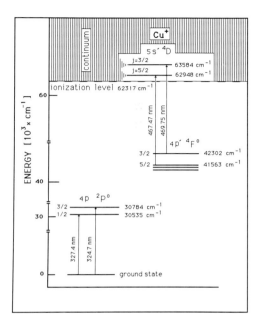

Figure 6 Transitions used to excite copper quartet autoionisation levels

parameter q which is the ratio of the matrix elements of the optical transition operators to the perturbed discrete level and to the continuum state. Fano introduced a reduced scale for the energy difference relative to the position of the unperturbed level by the following formula:

$$\varepsilon = [h(\omega - \omega_0)/2\pi + 1/q] \tag{28}$$

The Fano lineshape profile is described by :

$$\alpha(\varepsilon) = (q+\varepsilon)^2 / (1+\varepsilon^2). \tag{29}$$

3.3 Experimental results and discussion.

In order to reach the autoionisation levels we have tuned the dye laser to excite two of the quartet transitions $4p'\ ^4F^0_{3/2}$ -- $5s'\ ^4D_{3/2}$ at 469.75 nm and $4p'\ ^4F^0_{5/2}$ -- $5s'\ ^4D_{5/2}$ at 467.47 nm. There is a sharp voltage decrease at the initial time interval of the optogalvanic signal of autoionisation levels. This is demonstrated in Fig. 7 for the 5s' $^4D_{5/2}$ transition. A number of optogalvanic signals are shown. These are taken at different wavelengths across the autoionisation transition. The sharp decrease is the signature of the autoionisation OGS and is clearly different from that obtained from standard levels as seen in Fig 4. The sharp voltage fall at the beginning of the lower signal is due to the rapid decay of the autoionisation level. Every

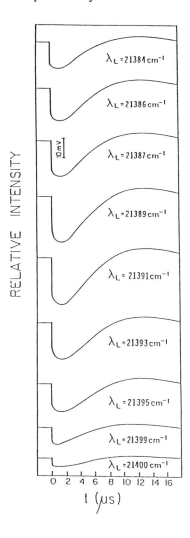

Figure 7 Optogalvanic signals obtained by scanning the line profile of the Cu autoionisation transitions $4p^{14}F^0_{5/2} - 5s'^4D_{5/2}$.

collision with a plasma electron can result in non-radiative decay which changes the excited atom from its autoionisation state to an ion in its final state. The largest optogalvanic signals were obtained when the the laser was in resonance with the transitions 4p' $^4F^o_{3/2}$ --5s' $^4D_{3/2}$ (at 469.75 nm) and 4p' $^4F^o_{5/2}$ -- 5s' $^4D_{5/2}$ (at 467.47nm). We scanned the dye laser wavelength across the resonance, each time measuring the optogalvanic signal (Fig 7). The wavelength change was in steps of 1 cm^{-1} as monitored by a double monochromator. The optogalvanic signals were measured and displayed as a function of the laser wavelength in Fig 8 for 4p' $^4F^o_{5/2}$ -- 5s' $^4D_{5/2}$ transition. This figure discribes the profile of the autoionisation level 5s' $^4D_{5/2}$.

We have used nonlinear least-square computer software *(Nelson 1983)* to fit the Fano asymmetric line profile based on equation (2.2) over the OGS data points. The best Fano parameters obtained for the level 5s' $^4D_{5/2}$ are q=-5, Γ =11 cm^{-1} and for the second level 5s' $^4D_{3/2}$ the computer fits gave q=4.15, Γ=10 cm^{-1}. For the level 5s' $^4D_{5/2}$ the variance is 3×10^{-3} and the

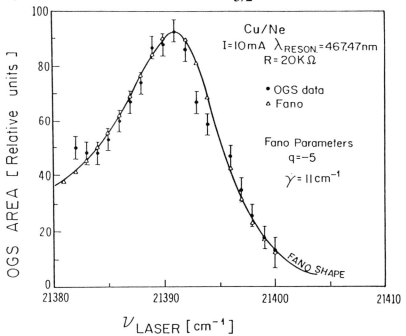

Figure 8 Optogalvanic line profile of the Cu autoionisation level 5s'$^4D_{5/2}$. The solid line is a computer fit of the Fano-type asymmetric line profile.

standard deviation is 0.054 and for the second level 5s' $^4D_{3/2}$ the variance is 7×10^{-4} and the standard deviation is 0.027. We would like to verify here that the experimental Fano assymetrical parameter relations of the two autoionisation transitions is:

$$|q(^4F^0_{5/2} -- 5s'\, ^4D_{5/2}) / q(^4F^0_{3/2} -- 5s'\, ^4D_{3/2})| = 1.20$$

We compare this result with that obtained from the multiplet intensity relations for these two transitions in pure LS Coupling which can be calculated from the tables of *White and Ellison (1933)*. This gives 1.3. The small deviation of the experimental value 1.2 from the theoretical value of 1.3 is due to the fact that the heavy copper atom is subjected to intermediate coupling (slight deviation from LS coupling toward jj coupling was observed earlier). There is good agreement between the theoretical and experimental q ratios. This makes the present method of using the optogalvanic effect to measure profiles of autoionisation levels a very attractive one. In our future work we will use this technique to study the doublet autoionisation levels of Cu and further levels of other elements.

4. Summary.

We have demonstrated the use of time resolved pulsed optogalvanic spectroscopy in studying plasma processes in discharges and in studying autoionising levels. In both cases the optogalvanic signals have their own specific signatures.

References

Allen C W, 1932, Phys. Rev. **39**, 55.
Assous R, 1973 Phys. Rev A, **7** 1213.
Ausschnitt C P, Bjorklund, G C and Freeman R, 1978, Appl. Phys. Lett., **33**, 851.
Ben Amar A, Shuker R and Erez G, 1981, Appl. Phys. Lett. **38**, (10) 763.
Ben Amar A, Shuker R and Erez G, 1983, J. Appl. Phys. **54**,3688.
Ben Amar A, Shuker R and Erez G, 1983, J. Appl. Phys. **54**, (7) 5685.
Cooper J, 1966, Reports on Progress in Physics, **29** 1 63.
C R C (Handbook of Chem and Phys) **52**nd, 1971, D-150.

Erez G, et al, 1979, IEEE J. Quantum Electron, **QE-15**, 1328.
Fano U, 1961, Phys. Rev. **124,** 1866.
Forrest L F, Pejsev V D, Smith K, Ross J and Wilson M,1987, J. Phys B **20** 3985.
Goldsmith, J E M, and Lawler J E, 1981, Contemp. Phys. **22**, 235, and refrences therein.
Kane D M, 1983, Opt. Com. **47**, (5) 317.
Kieffer L D and Dunn G H, 1966, Rev. Mod. Phys. **1**.
McDermott W and Nush C, 1975, Appl. Spectrosc. **29**, 408.
Moore C E, 1952, Atomic Energy Levels, Vol 2, NBS 35, US Department of Commerce.
Nelson D P and Homer L D, 1983, Non-linear Regression Analysis Naval Health Research Centre Report No **83-5.**
Pfaff J, Bergman M H and Saykally R J, 1984, Mol. Phys. **52**, 541.
Shenstone, A G, 1948, Phil. Trans. R. Soc A **241**, 297.
Shuker R, Ben Amar A and Erez G, 1983, J. Appl. Phys. **54** (10) 3688.
Striganov A R and Sventskii N S, 1968, Tables of spectral lines of neutral and ionized atoms (IFI / Plenum, New York).
White and Ellison, 1933, Phys. Rev. **44** 3985.
Wren D and Setser D, 1981, J. Chem. Phys. **74** (4) 2331.

Two-step laser optogalvanic spectroscopy of strontium: Even parity, J = 0–3, autoionizing 4dnl Rydberg states

A. Jimoyiannis, A. Bolovinos and P. Tsekeris
Atomic and Molecular Physics Laboratory, Physics Department, University of Ioannina, Ioannina, Greece

ABSTRACT: We have measured the energies of a few hundreds of even parity, autoionizing, J=0–3, 4dnl strontium Rydberg states using optogalvanic detection. These states are reached by a two–step pulsed dye laser excitation from the 4d5s metastables through the 4d5p $^3P_{0,1,2}$ intermediate states. The 4d5s states are populated by electronic collisions in a D.C. glow discharge through Sr plus He vapour in a heated cylindrical quartz cell.

1. INTRODUCTION

The level structure and properties of two-electron systems are of particular interest for understanding many atoms. Alkaline earths are well suited for such studies, since they have two electrons outside closed shells. The availability of tunable dye lasers and the development of sensitive detection techniques have greatly increased the information on these atoms beyond the J=1 odd-parity Rydberg states that can be detected by one photon absorption (Garton et al 1968). Thus the bound spectra of these atoms have been extensively studied in the previous decade, while in this decade the interest has moved to their autoionizing states (Aymar 1984).

Up to now the only autoionized states observed in Sr are the odd 4dnl (Garton et al 1968), 5pns (Xu et al 1986) J=1 and 4dnl (Kompitsas et al 1990), 5pnd (Xu et al 1987) J=3 ones. In this work we report on the observation of several 4dnl even parity Sr autoionizing Rydberg series. We detect these series in a d.c. glow discharge by a two-step laser optogalvanic technique. In the discharge the Sr even parity metastable states 4d5s, which are ~ 18.000 cm^{-1} above the ground $5s^2$ $^1S_0$ state (Figure 1), are sufficiently populated by electronic collisions. From these states we can excite in the first step the odd parity 4d5p states with J=0-4, using laser light in the green-red spectral range. In the second step we reach the 4dns, 4dnd and 4dng even parity autoionizing states with J=0-5, using laser light in the violet-green spectral range.

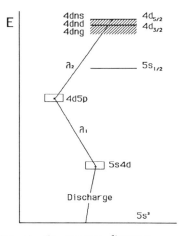

Fig. 1. Sr energy diagram.

The advantage of this technique is that we can excite high energy and high J Rydberg states using only two easily available laser wavelengths. If we had to start from the ground J=0 level, we would need up to five steps to reach the same states and this would introduce complexity in the experimental set-up and the analysis of the results. On the other hand the discharge environment restricts our resolution making the detection of principal quantum numbers n higher than ~35 difficult.

In this work we present results in the energy region above ~ 59000 cm^{-1} up to the $4d_{5/2}$ ionization limit. They are obtained through the excitation of the 4d5p $^3P_{0,1,2}$ intermediates, which enables us to detect states with J=0-3 values. The J=4,5 states are presently under investigation using the 4d5p $^3F_1$, $^3F_4$ intermediates.

2. EXPERIMENTAL METHOD

The experimental method is the same with that used by Camus et al (1979) in their study of barium atoms. The set-up shown in Figure 2 includes two pulsed dye lasers pumped by the same 7 nsec, 3mJ N$_2$-laser (SOPRA 804C), a heated cell, an evacuated home made Fabry-Perot etalon with a free spectral range of 1.841±0.001 cm^{-1} followed by a fast photodiode (MRD 500), two gated integrators (PAR 165/164), for recording the optogalvanic and the photodiode signal, and a dual pen chart-recorder.

The heated cell is similar to that of Camus (1974). It is a 30 cm long quartz tube of 15 mm diameter with two removable electrodes from stainless steel (anode) and nickel (cathode). The heated region has a length of ~ 15 cm and includes a thin stainless steel tube with a stainless steel mesh as a wick. When filled with Ba, this cell can operate as a heat-pipe too, but we found that this is not possible with Sr. So we need usually to refill the cell with Sr after ~20 hours of operation.

The discharge is produced by a well stabilised HV power supply (OLTRONIX A3K4-40R) and the optimum operation conditions are 600-700 V for total voltage, 12-15 mA, 2-4 Torr of total gas pressure (He & Sr vapour) and a temperature of ~700 °C, which corresponds to ~1 Torr of Sr vapour (Nesmeyanov 1963).

We used a home-made dye laser oscillator (5 nsec, 10-20 µJ) for the first excitation step at wavelength λ_1 and a commercial dye laser oscillator-amplifier (5 nsec, 50-100 µJ) (SOPRA LCR I) for the second step at λ_2. Both dye lasers have a linewidth of ~0.4 cm^{-1}. Under these conditions the damped oscillating OG signal has peak amplitudes up to a few volts, while the OG signal of the first excitation step is 10-50 mV only. The discharge noise level is no more than 5 mV.

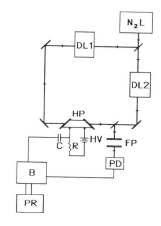

Fig. 2. Experimental set-up: N$_2$-L: Nitrogen laser; DL: dye lasers; HC: heated cell; HV: power supply; R: ballast resistor 22 K; C: coupling capacitor 2.2 nF; FP: Fabry-Perot etalon; PD: photodiode; B: Boxcar averager; PR: pen recorder.

3. RESULTS AND DISCUSSION

In Figure 3a we show a typical spectrum, taken when both laser beams are interacting with the atomic gas, and the Fabry-Perot interference fringes detected by the fast photodiode. Figure 3b shows the spectrum recorded when the first laser beam is blocked off. The lines seen in this case are due to transitions between known bound Sr states (Rubbmark and Borgström 1978) and are used as calibration standards for calculating the energies of the 4d5p→4dnl transitions, which are the additional lines seen in Figure 3a. The energies of the 4dnl states are found by adding these transition energies to the known energy of the particular 4d5p intermediate state. The accuracy of our results is estimated to be ±0.1 cm^{-1} for sharp lines and 2 or 3 times worse for broader and/or weaker lines.

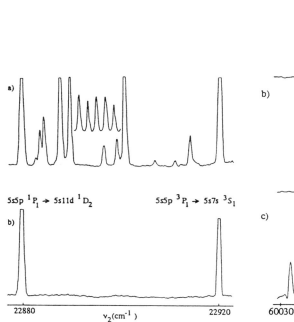

Fig. 3. a) Two-step optogalvanic spectrum of Sr. A part of interference fringe pattern is included. b) Same spectrum without first step excitation.

Fig. 4. Two-step optogalvanic spectrum of Sr through the intermediates 4d5p a) $^3P_0$, b) $^3P_1$, c) $^3P_2$.

Figure 4 shows a typical spectral region reached through the different ^3P_J intermediate levels. Applying the $\Delta J = 0, \pm 1$ (J=0 $\leftarrow\!\!\!/\!\!\!\rightarrow$ J=0) selection rule, the J values of the 4dnl states are easily derived. Those reached from $^3P_0$ have J=1 only, while through $^3P_1$ and $^3P_2$ we reach states with J=0-2 and J=1-3 respectively. Thus the states seen in all three spectra will have J=1, the ones seen only in the spectra recorded by exciting the $^3P_1$ and $^3P_2$ intermediates have J=2 and the ones seen only through the $^3P_1$ or $^3P_2$ have J=0 or J=3 respectively. We are thus able to deduce the J values of the vast majority of the observed states. Their energy values will be published elsewhere. Most of the detected levels can be assigned to series of the same J and a specific configuration with constant or smoothly varying n*(mod 1) values. For the remaining levels, which are all found below the $4d_{3/2}$ limit, we need theoretical calculations, like the MQDT and R-matrix methods (Kompitsas et al 1990), to determine their configurations. In this second group belong the J=0 levels too, which correspond mostly to weak signals.

TABLE I. J = 0-3 even 4dnl Sr configurations

Configuration	J (expected)	J (observed)
$4d_{3/2}$ ns	1, 2	1
$4d_{3/2}$ nd	0, 1*, 2*, 3*	1*, 2, 3
$4d_{3/2}$ ng	2, 3*	2, 3 (not resolved)
$4d_{5/2}$ ns	2, 3	2, 3
$4d_{5/2}$ nd	0, 1*, 2*, 3*	1*, 2*, 3*
$4d_{5/2}$ ng	1, 2*, 3*	1, 2, 3 (not resolved)

The asterisk denotes two series with the same J value.

In table I we give the number of the expected and observed J=0-3 even 4dnl configurations, while in figure 5 we present the series with constant n*(mod 1) converging to the $4d_{3/2}$ and $4d_{5/2}$ ionization limits. There is a triad of

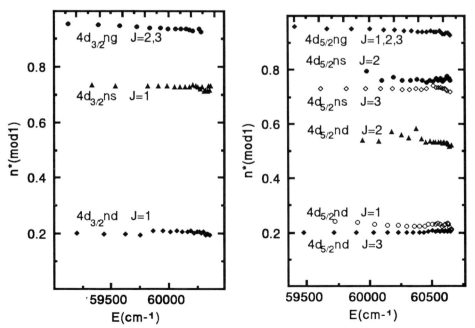

Fig. 5. n*(mod 1) values for well behaved Sr Rydberg series converging to a) $4d_{3/2}$ and b) $4d_{5/2}$ ionization limits.

4dnd series with J=1,2,3 converging to each 4d limit, whose low members are well resolved, while the members above $n^* \sim 25$ for $4d_{3/2}$ and $n^* \sim 17$ for $4d_{5/2}$ are not. As far as the 4dng series are concerned, their different J states cannot be resolved, as expected, because of the small fine structure of the g electrons.

Aknowledgements: We thank Prof. P. Camus for his help in the various stages of this work and Mr G. Skalistis for his assistance with electronics.

REFERENCES

Aymar M 1984 *Phys. Rep.* **110** 163
Camus P 1974 *J. Phys.* **B7** 1154
Camus P, Dieulin M and Morillon C 1979 *J. Phys. Lett.* **40** L513
Garton W R S , Grasdalen G L, Parkinson W H and Reeves E M 1968 *J. Phys.* **B1** 106
Kompitsas M, Cohen S, Nicolaides C A, Robaux O, Aymar M and Camus P 1990 *J. Phys.* **B23** 2247
Nesmeyanov A N 1963 *Vapor Pressure of the Chemical Elements* (New York: Elsevier Publishing Co.)
Rubbmark J R and Borgström S A 1978 *Phys. Scripta* **18** 196
Xu E Y, Zhu Y, Mullins O C and Gallagher T F 1986 *Phys. Rev.* **A33** 2401
Xu E Y, Zhu Y, Mullins O C and Gallagher T F 1987 *Phys. Rev.* **A35** 1138

High resolution IMOGS measurements in heavy elements

D Ashkenasi, G Klemz and H-D Kronfeldt
Optisches Institut, TU Berlin, Straβe des 17. Juni 135, 1000 Berlin 12, Germany

ABSTRACT

We discuss and demonstrate the application of high resolution intermodulated optogalvanic saturation (IMOGS) spectroscopy on lanthanide elements such as Eu, Er and Gd for investigations of higher order atomic effects.

1 INTRODUCTION

For the determination of fundamental atomic data such as hyperfine structure splittings (hfs) or isotope shifts (IS) in atoms and ions many different high resolution laser spectroscopic methods have been proposed and developed; eg a good summary can be found in Demtröder (1988). Among these, optogalvanic detection (first discovered by Penning (1928)), in combination with tunable dye lasers, represents a powerful technique for high resolution atomic and molecular spectroscopy, eg Shimoda (1976), Goldsmith et al (1979) and Lawler et al (1979). After the first optogalvanic spectroscopy on alkaline and alkaline earth atoms (eg Green et al (1976)), an increasing interest developed in using optogalvanic detection in many fields. Optogalvanic detection provides a convenient means of observing an optical transition without requiring any conventional optical detectors; the discharge combines both the spectroscopic sample and the detector. An important advantage of optogalvanic detection is that in a hollow cathode (hc) discharge virtually all the energy levels of the neutral atoms are populated, especially the metastable levels. Many transitions in ions are detectable using the optogalvanic method. It represents an important alternative to absorption and fluorescence techniques. An extensive bibliography of optogalvanic spectroscopy, covering the 1923–1983 period, was compiled by Camus (1983). We employed the high resolution technique of intermodulated optogalvanic saturation (IMOGS), a

© 1991 IOP Publishing Ltd

form of saturation spectroscopy, for investigations of heavy elements, e.g. the lanthanides.

2 EXPERIMENT

The experimental set-up for the high resolution Doppler-reduced IMOGS measurements is shown schematically in figure 1. The output beam (output power: 200-600mW) of a narrow band tunable light source – a Coherent ring dye laser system – is split into two beams of roughly equal intensity. The two counter-propagating light beams overlapping in the hc discharge are mechanically chopped with different frequency f_1 and f_2 while the lock-in amplifier system is tuned to the sum frequency $f_1 + f_2$, ie the intermodulated mode. A cavity marker with a free spectral range of 149.56(14) MHz, permits highly accurate frequency calibration of the detected spectra.

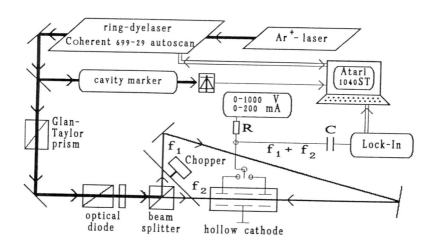

Fig 1 Experimental set-up for high resolution IMOGS measurements

The see-through LN_2-cooled hc used in the IMOGS measurements, described in detail by Seifert et al (1989), is conveniently demountable. The cathode, an exchangeable copper block, is filled with a thin foil of the lanthanide material under investigation: Eu, Gd or Er. The hc was constructed symmetrically with adjustable anode–cathode distances. All surfaces in the discharge region were extremely well polished to avoid any possible plasma instabilities and to minimize discharge noise. We obtained the best results (low noise) with pure Kr(99.99 vol%) as the discharge gas, with a pressure around 1mbar and with a discharge current of 10 to about 100mA. Special care was required during preparation of the hc detector and a burn–in period with several gas changes was necessary to achieve optimum discharge performance.

3 OPTOGALVANIC FINE STRUCTURE SCANS

Before starting high resolution IMOGS investigations in the lanthanide element series and before the selection of appropriate transitions, in many cases it was necessary to study fine structure scans of chosen wavelength intervals. Optogalvanic signal intensities can differ strongly from optical emission intensities although theoretically there is a correspondence between the laser–induced voltage change and the intensity of the emission line as discussed and demonstrated for uranium by Keller *et al* (1980). Figure 2 is an example of a large optogalvanic fine structure scan covering about 25nm of Eu, together with an optogalvanic Ne reference spectrum for wavelength calibration, obtained using a pulsed dye laser system. The Eu lines are identified; the other lines originated from the discharge gas Kr. The Eu II lines all produce a negative discharge signal.

4 IMOGS INVESTIGATIONS IN Eu

We used the set–up shown in Figure 1 for the IMOGS measurements. Fig 3 is a typical example of a high resolution IMOGS measurement on the Eu I line 584.58nm. Single hfs components are clearly resolved but special care must be taken to identify the cross–over signals due to saturation effects. The IMOGS signal shown in Fig 4 is a more complicated case of the double–line at 594.27nm. The high resolution spectrum was sufficient to determine the hfs and IS with high accuracy, eg to evaluate crossed–second order effects in the isotope shift. For a more extensive view on the hfs and IS interpretation resulting from Eu IMOGS measurements we refer to our paper, Seifert *et al* (1989).

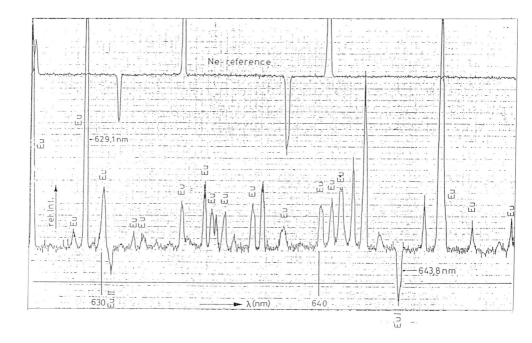

Fig 2 Optogalvanic Eu fine structure recording from 625–650 nm with Ne reference

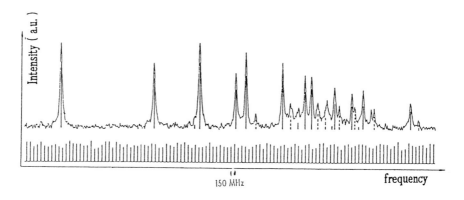

Fig 3 IMOGS intensity profile of the Eu I line 584.58nm (- - - -: cross-over signals) transition ($4f^76s7p$ v $^8P_{3/2}$ − $4f^75d6sb$ $^8D_{1/2}$)

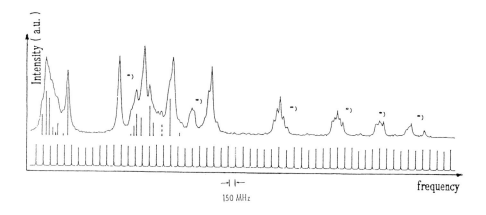

Fig 4　IMOGS intensity profile of the EuI double-line 594.27nm
(- -:cross-over signals)

transition:　: $4f^75d6p$ y $^8D_{3/2}$ - $4f^75d6s$ b $^8D_{5/2}$

transition*):　$4f^75d6p$ z $^6F_{5/2}$ - $4f^75d6s$ a $^6D_{5/2}$

5 IMOGS INVESTIGATION OF Gd AND Er

While high resolution IMOGS measurements on Eu in our home-made hc were fairly straightforward, we experienced some difficulty achieving a good optogalvanic signal in Gd and Er, due to strong non-periodic spiking. To improve the signal to noise ratio and establish conditions necessary for high spectral resolution work, we tried a bulk cathode made completely of the lanthanide material instead of a foil. It should be noted that with the commercial Hamamatsu see-through hc "galvatron" it is possible to obtain reliable optogalvanic detection for Er and Gd. Fig 5 shows the IMOGS intensity profile of the Gd I 569.62nm line using the Gd "galvatron". The even isotopes ^{160}Gd, ^{158}Gd and ^{156}Gd are clearly recognizable while the hfs splitting of the ^{155}Gd and ^{157}Gd is largely unresolved because of the Doppler-broadened background. A similar situation is seen in the Er I 575.25nm line, as shown in Fig 6, where the line profiles of only the even Er isotopes are observable. However, hfs and IS data are obtainable in these cases. One great disadvantage of the commercial hc is the lack of cooling which prevents us from running the discharge current above 10 - 15mA without substantially reducing the tube lifetime.

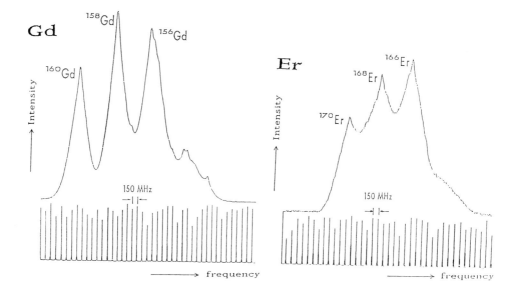

Fig 5   IMOGS intensity profile of the Gd I 569.62nm line

Fig 6   IMOGS intensity profile of the Er I 575.25nm line

6 CONCLUSION

High resolution IMOGS measurements on Eu, ie under 80MHz resolution in a 20GHz scan, on Eu are straightforward to make.   This permits investigations of even second order atomic effects, eg in the hfs or IS; in contrast to a general conclusion by Bachov et al (1982).   High resolution IMOGS measurements on Er and Gd are more difficult because of discharge noise. Further effort in conditioning the discharge to a reliable low noise level will be necessary, but is most promising.

ACKNOWLEDGMENT

We are indebted to Dr P Seifert (former member of our group) for the Eu experiments.

REFERENCES

Bachov H M, Manson P J and Sandeman R J 1982, Optics Comm. **43** 5.

Camus P (Editor) 1983, Proceedings of the International Colloquium on Optogalvanic Spectroscopy and its Applications **44** C7 (France: Journal de Physique).

Demtröder W 1988, Laser Spectroscopy – Basic Concepts and Instrumentation 3rd Printing (Berlin: Springer).

Goldsmith J E M, Ferguson A I, Lawler J E and Schawlow A L 1979, Opt. Lett. **4** 230.

Green R B, Keller R A Luther G G, Schenk P K and Travis J C 1976, Appl. Phys. Lett. **29** 727.

Keller R A, Engleman R Jr. and Palmer B A 1980, Appl. Optics **19** 6.

Lawler J E, Ferguson A I, Goldsmith J E M, Jackson D J and Schawlow A L 1979, Phys. Rev. Lett. **42** 1046.

Penning F M 1928 Physica **8** 137.

Seifert P, Weber D-J, Kronfeldt H-D and Winkler R 1989, Z Phys. D **14** 99.

Shimoda K (Editor) 1976 High-Resolution Laser Spectroscopy (Berlin: Springer).

An original method for measuring the conductivity of low pressure discharges

R.I.Cherry, T.R.Robinson

Dept. of Physics and Astronomy, University of Leicester, Leicester, LE1 7HR

ABSTRACT: A new method of measuring plasma conductivity based at radio frequencies is presented. The application and merits of the crystal based conductivity probe (XBCP) are discussed. Example form factors relating the conductivity of the plasma to the geometry of a typical device are given. From these form factors general properties of XBCP's are inferred. The results from an investigation into an Argon/Mercury discharge using a XBCP are given.

1. INTRODUCTION

The detection of the optogalvanic effect is usually at its most sensitive and simplest when global changes in the plasma conductivity are inferred from the change in discharge maintainance current. A global measurement does however have disadvantages. This is particularly evident when it comes to interpreting the macroscopic signal in terms of the myriad microscopic processes of which it is composed. There may therefore be some advantage in having a technique which offers a means of making local plasma conductivity measurements

Although numerous schemes have been suggested for the measurement of local plasma conductivity using radio frequencies (Heald *et al* (1961), Savic *et al* (1962), Bandelier *et al* (1988)) no one method has become established. Many of these methods are based on the loading of a simple LC resonant circuit the inductor of which is wound concentrically about the discharge tube. It is assumed that a mutual inductance is formed between the coil and the plasma and the resistive part of the latter loads the resonant circuit which then becomes 'de-Qed'. There are at least three major limitations to these techniques. Firstly, a circuit which is sensitive to its wound components can suffer from drifts associated with temperature and humidity. This situation is exacerbated when the coil is in close proximity to a heat source such as a plasma. Secondly, the Q of LC cicuits is quite small, this can lead to substantial errors in the absolute measurement of Q in some circuit configurations (Cherry 1989). Finally although conductivity probes have acheived some degree of success as diagnostics for high conductivity plasmas such as shocks and arcs, in the low pressure and low conductivity case there is significant evidence to suggest that it is in fact the electric field of the coil, rather than the magnetic field, which is the responsible coupling mechanism (Cherry 1989, Ghosal 1976).

Conventional means of measuring plasma conductivity rely on the effect the impedance presented by the plasma has on an external circuit. Generally a low pressure discharge coupled to a circuit by an electric field will add an

© 1991 IOP Publishing Ltd

impedance composed of a capacitive (coupling) part in series with a resistive (plasma conductivity) part. There are of course more sophisticated methods available nowdays for measuring complex impedances. Network analysers might appear to offer a means of determining this impedance however the utility of these devices in this instance is limited as the phase measurements which are required to distinguish betweeen the real and imaginary parts of the impedance can not be made with sufficient precision.

The method we propose for making plasma conductivity measurements in low pressure discharges dispenses completely with the inductive component that is usually associated with circuits devised for making this kind of measurement. In this method the impedance presented by the discharge is coupled via the electric field of an air gap capacitor which is in turn connected in series with a commercially available resonant crystal (Fig 1.). It is from the changes induced in this network by the presence of the plasma that the plasma conductivity is inferred. Apart from the simplicity of the system, there is an inherent robustness and stability in operation which arises from the use of the crystal.

2. THE XBCP

Figure 1 illustrates the experimental configuration and the accepted equivalent network for the crystal based conductivty probe (XBCP). The resonant crystal is housed in an oven maintained at 51°C. A signal generator (resolution of 1Hz) and a true RMS voltmeter is used to sweep out the transfer function of the network. From this response the selectivity (Q) and resonant frequency (Fr) is determined. The effect of a plasma between the plates is to change the Q and Fr of the circuit. The reactive part of the impedance (Cp) is the principle cause of the shift in frequency. This shift can be measured with a high degree of precision thereby allowing the reactive component to be determined.

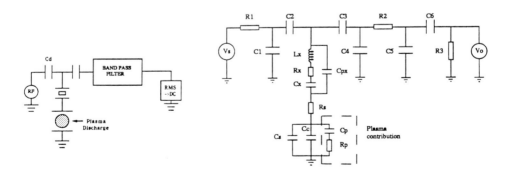

Fig 1: a) Arrangement for measuring transfer function of XBCP
       b) Equivalent circuit

Although the XBCP is intrisically a band-pass device, it has proved useful to attenuate excessive low frequency pickup arising from mains hum and the heating element of the oven. A high frequency limit is desirable if the XBCP is being applied to a radio frequency sustained plasma. For these reasons a simple first order band pass filter (3dB at 100kHz and 70 MHz) forms the final stage in the device (Fig 1.).

The components that make up the equivalent network can all be measured using conventional bridges, or, in the case of the resonant crystal, a network analyser. Although rough measurements of Rs and Cs are possible it is expedient to reconcile the calculated Q and Fr with the measured Q and Fr of the system by adjusting Rs and Cs. In this manner the ambient parameters of the network can be set. The presence of the plasma, contained in a glass tube (1.3cm inner diameter) adds an equivalent impedance represented by Rp and Cp. This impedance causes both a shift in frequency and a change in the Q of the network.It is not possible to reduce these changes to an analytical relationship to Rp and Cp. Both setting the ambient component values and the effect of the plasma impedance has to be modelled numerically (Cherry 1990a).

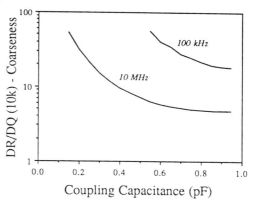

Fig 2: Coarseness of XBCP at 100kHz & 10MHz

3. NETWORK ANALYSIS

The change in the ambient parameters is predicted through calculating the voltage transfer function of Fig 1. Such an analysis is easily extended to allow analysis of the sensitivity of the network to changes over a wide range of Rp and Cp values.

Although resonant crystals can be obtained that cover a wide frequency range (10kHz<Fr<500MHz) there are limitations to the frequency that can be used. Fig 2 is a graph of the coarseness (the reciprocal of sensitivity) as calculated for a set of component values which are typical to a 10 MHZ and 100kHz XBCP. The network analysis indicates that for $1k\Omega<Rp<20k\Omega$, the change in Q, for any one coupling capacitance is proportional to the change in Rp - an observation which allows us to express the coarseness as a function of the two dominant variables; coupling capacitance and frequency. The plot indicates that 10 MHz is about the optimum frequency for XBCP operation. Above this frequency there is a danger that the skin depth in the plasma will cause complications and below this the sensitivity is much reduced.

4. LAPLACE MODELLING OF FRONT END OF XBCP

The gap which forms the air gap capacitor between which the discharge is placed is designed to allow the fields interacting with the plasma to be calculated as accurately as possible (Cherry et al 1990b). In particular cylindrical geometry is imposed on the plates forming the capacitor and the electrodes are housed between two earthed planes which further define the boundary conditions. Laplaces equation can be solved for this geometry and from a consideration of the power dissipated in the plasma (assumed to be a conductor exhibiting a predefined conductivity profile) an expression relating the resistance Rp measured by the XBCP to the peak conductivity can be derived:

$$Rp = 1/[\sigma_{peak} * \int f(r,z,D) \, dV * F(x,y,z)] \tag{1}$$

where $F(x,y,D)$ contains the field information and $f(x,y,z)$ is a function that represents the plasma conductivity profile. Fig 3. shows the calculated form of this equation for three plasma conductivity profiles as a fuction of the parameter D (the electrode diameter). Unfortunately the dielectric nature of the glass tube containing the plasma can not readily be incorporated in the solution as there is a conflict of geometries. However, a number of observations follow from solving the equation for the non-dielectric case.

Figs 2 and 3 give some indication of the sensitivity of the device to plasma conditions. It is possible to detect changes in Rp of $\approx 100\Omega$, this suggests that in principle quite subtle changes in plasma conductivity can be monitored, although equation 1 must be solved for any one XBCP if detailed information is required.

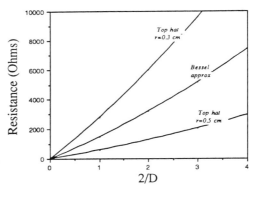

Fig 3 :Theoretical variation of resistance presented to XBCP as a function of D for three conductivity profiles

It is useful to establish the range of application of the device. This range is determined by two factors. Firstly the applied frequency of the XBCP must be much less than the electron-neutral collision frequency of the plasma for the plasma to behave as a conductor. However Fr must also be small enough so the skin depth in the plasma is unimportant. Secondly, if the electrodes of a general XBCP are taken to be cylindrical and it is assumed that the majority of the power is dissipated in the plasma within a region that is equivalent in dimension to the diameter (D=2r) of the electrode, then, as the network analysis indicated that the network is most sensitive to detecting Rp$\approx$10kΩ a simple analytical argument shows that r is optimum in detecting a plasma conductivity σ (SI units):

$$\sigma_{(average \, over \, r)} = \{10^{-3}\}/r \tag{2}$$

The average conductivity over the distance r can easily fall short of the peak conductivity of the plasma in a typical positive column discharge by up to an order of magnitude. As Fig 3 suggests the XBCP is particularly sensitive to the edge profile of the plasma and as will be shown later it is useful to have a design of XBCP which allows the aspect D of the probe to be varied.

5. APPLICATION AND VALIDITY OF METHOD

A number of theoretical descriptions exist for Argon mercury dischages (Hirsh 1978) of these one of the most explicit is that of Cayless (1962). This description has been implemented (Lister 1989) and subsequently extendeded to demonstrate the relationship between the power maintaining the discharge and the resulting conductivity.

FIG 4a: Theoretical variation of average plasma conductivity as a function of power dissipated in an Ar/Hg discharge. Legends indicate pressure of Argon. Mercury pressure fixed at 6.7 Microns. Cold spot at 40 C

Fig 4a depicts the behaviour of the plasma conductivity as a function of pressure and applied power. Three different background pressures of argon are shown and the cold spot of the discharge is fixed at 40°C. Fig 4b shows the results obtained from an early design of XBCP (one without temperature stabilization or well defined fields). Although it was not possible at this stage to directly relate the measured Rp to the peak conductivity, it is clear that for the 1 Torr and 5 Torr case the XBCP could demonstrate the predicited trends in plasma behaviour. With the 10 Torr plasma the network analysis began to faulter - the interpretation relies on cross plotting and in this instance this process fell on an ill-conditioned part of the curves - hence the large error bars. This latter observation has since led to a more versatile design of XBCP where the radius of the electrode can be readily changed in accordance with equation 2.

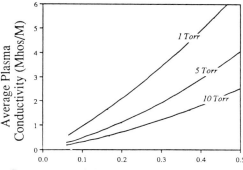

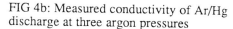

FIG 4b: Measured conductivity of Ar/Hg discharge at three argon pressures

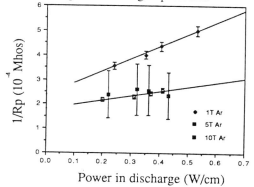

6. APPLICATION TO OGS

Although there are considerable advantages in sensitivity in restricting optogalvanic studies to the sheath regions of the cathode, there may be substantial gains to be made in understanding when monitoring the effect in the positive column, particularly as the interaction between the XBCP and the plasma appears to lend itself to straightforward modelling. Furthermore, the XBCP has been shown (Cherry 1990) to be equally applicable to electrodeless discharges which has implications for optogalvanic spectroscopy (A.Sasso 1990).

7. SUMMARY

As the electrodes of a XBCP are mounted outside the discharge vessel, the diagnostic is both flexible and non-intrusive. These features, combined with the comparatively local nature of the measurement, suggest a number of possible applications of the diagnostic both as a stand alone device and in the field of optogalvanic spectroscopy.

8. ACKNOWLEDGEMENTS

The authors would like to thank G.Lister for his helpful discussions and loan of his implementation of the Cayless diffusion code.

Bandelier B, Pointu A M, Rioux-Damidau Zeller P 1988 *IEEE Trans. Mags.* 88,24,1
Cayless M A 1963 *Brit.J.Appl.Phys. 14,863*
Cherry R 1989 *Thorn Lighting Internal Report*
Cherry R, Robinson T 1990a *Rev. Sci. Instru. (In press)*
Cherry R, Robinson T 1990b *Rev. Sci. Instru. (in press)*
Ghosal S K, Nandi G P, Sen S N *Int. J. Elecs. 509,41,5,1976*
Heald M A , Wharton C B 1965 *Plasma Diagnostics with Microwaves (New York: Wiley)*
Hirsh N M, Oskam H J 1978 *Gaseous Electronics Vol 1 (San Francisco: Academic press)*
Lister G 1989 *Thorn Lighting Internal Report RM 1795*
Sasso A 1990 *This conference*
Savic P, Boult G T 1964 *J.Sci.Instru. 258,39*

Ultra-trace analysis of NO by high resolution laser fluorescence and ionization spectroscopy

M Hippler, A J Yates and J Pfab

Heriot-Watt University, Edinburgh

ABSTRACT: A new laser technique for the reliable ultra-trace analysis of NO is presented. The combination of free jet expansion and LIF or REMPI detection combines spectral selectivity with high sensitivity. Current detection limits are 5 ppt of NO in 1 atm of Argon.

1. INTRODUCTION

NO (nitric oxide) is formed in the atmosphere by photo-dissociation of NO_2 into NO and O and by all combustion processes. The formation of ozone and the degradation of organic compounds depend critically on trace levels of NO. Reliable ultra-trace analysis techniques are therefore needed for a better understanding of atmospheric chemistry. For polluted air standard analytical methods such as chemi-luminescence are well established. These techniques, however, can only be extended to the trace analysis of NO in the sub-ppb range with great expense. The reliability of the results is debatable due mainly to interference from other atmospheric constituents. A further intercomparison (see, e.g. Hoell et al. 1987) with instruments using alternative detection techniques is urgently required.
We present here preliminary results obtained with new laser techniques for the reliable detection of NO in the ppt range.

2. EXPERIMENTAL

The experimental set-up consists of a vacuum chamber and a laser system. Mixtures of known concentrations of NO in Argon were prepared on a vacuum line. A piezoelectric pulsed nozzle (Proch, Trickl 1989) provides gas pulses of 150 μs duration with a repetition rate of 10Hz into the vacuum chamber. With a backing pressure of 1 atm and a chamber pressure < 10^{-4} torr, rotational cooling of the NO molecules to T < 5K is easily achieved. Different nozzle orifices have been investigated (see next section). After a suitable delay, the output of an excimer laser pumped dye laser (EMG50 and FL2001 or EMG101MSC and FL3002, Lambda Physik) is focussed with a 50 cm focal length lens into the vacuum chamber. For the experiments with 226nm laser light, the dye laser output is frequency doubled with a BBO- crystal and directed into the chamber without focussing.

© 1991 IOP Publishing Ltd

Laser-induced fluorescence (LIF) is collected by a 10cm focal
length lens and imaged onto a UV-sensitive photomultiplier.
Alternatively, ions generated by multiple photon ionization
(REMPI) are collected by an assembly of two parallel plates,
spaced by 3cm and biased by 200V dc. The small ion current is
amplified by a current to voltage amplifier (Adams et al.1980).
Signals of both LIF and REMPI experiments are stored via a
boxcar integrator and an interface in a computer for further
data processing. Signal from a reference cell filled with NO
traces in Argon at room temperature is taken to compensate for
laser power fluctuations and variations caused by the limited
lifetime of the laser dyes. An artist's view of the LIF
experimental set-up is shown in fig.1.

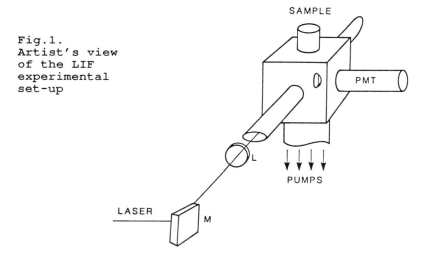

Fig.1. Artist's view of the LIF experimental set-up

3. RESULTS AND DISCUSSION

As a result of supersonic jet cooling, the NO molecules will
preferably occupy their lowest rotational quantum states in
contrast to room temperature. At a typical temperature of 4K,
75% of the NO molecules are in the lowest rotational quantum
state. Here, they are accessible for laser interaction at a
single resonant transition, which enhances the signal strength.
Fig.2 shows a comparison of a room temperature and a ultra-cold
spectrum of NO. The dramatic simplification of the spectrum
through cooling is obvious. The NO concentration can be
inferred from the intensity of a single well defined spectral
transition. This results in extremely high spectral selectivity
and makes interferences from other atmospheric constituents
improbable.
Cooling becomes more efficient with a decrease in the size of
the orifice and an increase in the distance between nozzle and
probing region. On the other hand, the signal strength
increases with the number density of NO molecules in the
probing region. The number density increases with larger nozzle
orifices and smaller probing distances. Optimum conditions
found are a 0.5mm nozzle orifice and a probing distance of 5mm.

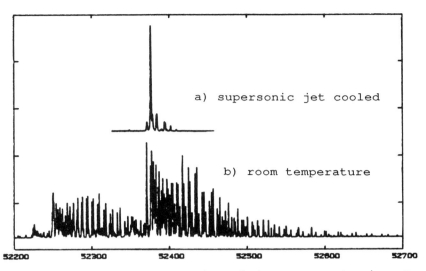

Fig.2. Comparison of room temperature and expansion cooled NO REMPI spectrum ($C\,^2\pi \leftarrow X\,^2\pi$ 2+1 REMPI)

In table 1, detection limits for the LIF experiments are summarised. As expected, 1-photon LIF is most sensitive, because the cross-section for a 1-photon process is orders of magnitude larger than for a 2-photon process.

Table 1. Detection limits for NO

method	nozzle	detection limit
2-photon LIF	0.05 mm	1 ppm
2-photon LIF	0.50 mm	5 ppb
1-photon LIF	0.05 mm	0.4 ppb
1-photon LIF	0.50 mm	5 ppt

For the REMPI experiments, preliminary results indicate a similar sensitivity to LIF. Detection limits below 50 ppb with (2+1) REMPI via the C-state and in the ppt range with (1+1) REMPI via the A-state are anticipated.
In fig.3 the detection schemes employed via the different electronic states of NO are summarised.
Problems to be taken into account are background effects and outgassing of rubber and teflon components when previously exposed to high NO concentrations. Furthermore, other NO containing molecules like NO_2 or CH_3ONO can be photofragmented to produce hot NO, especially when using focussed laser light. But as evident from fig.2, hot NO can be discriminated readily from ultra-cold NO. In addition, the hot NO from photodissociation is distributed over a wide range of quantum states leading to a drastic reduction in the intensity. A study of the photodissociation of alkyl nitrites is in preparation (Hippler,

Al-Janabi, Pfab 1991). Further work will be carried out to
establish and test procedures for future field measurements
using calibration techniques with NO samples prepared from
certified gas mixtures.

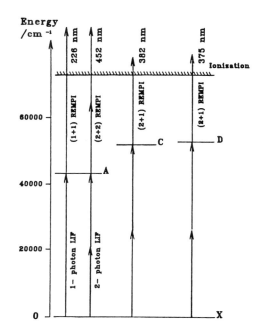

Fig.3. Detection schemes via the different electronic states of NO

4. CONCLUSIONS

Supersonic jet LIF and REMPI constitute highly promising
techniques for the analysis of NO in the important sub ppb-
range. The techniques combine high sensitivity with high
spectral selectivity and freedom from interference of other
atmospheric constituents. Optimally designed equipment would be
suitable for mobile measuring stations in vans or small
aircraft, with costs comparable to other ultra-trace analysis
techniques. The analysis of trace pollutants other than NO may
be also feasible using the same techniques.

5. ACKNOWLEDGEMENTS

We are grateful to the Commission of the European Community for
funding this work (project B-01-46) and the UK Science and
Engineering Research Concil for support.

6. REFERENCES

Adams T E, Morrison R J S and Grant E R 1980
  Rev.Sci.Instr. 51 141
Hippler M, Al-Janabi F and Pfab J 1991 in preparation
Hoell J M, Gregory G, Davies D D, Ridley B A et al. 1987
  J.Geophys.Res. 92 1995
Proch D, Trickl T 1989 Rev.Sci.Instr. 60 713

Time-involved fluorescence studies of the laser dye DCM

G. Hungerford, D.J.S. Birch and R.E. Imhof,
Department of Physics and Applied Physics, University of Strathclyde, Glasgow, UK.

Abstract

We report the fluorescence decay behaviour of the laser dye DCM in both methanol and dimethyl sulphoxide using the single photon counting technique. In methanol we confirm that the decay is best described using a biexponential decay model with the most likely origins of the fluorescence being the cis and trans isomeric forms of the molecule. In dimethyl sulphoxide we observe a deviation from the previously reported monoexponential decay model. Here we find that the fluorescence is best described using either a biexponential model or one involving a distribution of lifetimes about a mean and this we interpret as providing the first evidence that both cis and trans emissions might be occurring in DMSO.

1. Introduction

DCM (4-dicyano-methylene-2-methyl-6-p-dimethylaminostiryl-4H-pyran) is widely used as a laser dye because it exhibits a wide tuning range and high conversion efficiency (Hammond 1979). It also features a large shift between absorption and fluorescence maxima and its stability has been reported as reasonably good (Antonov and Hohla 1983). As well as being employed as a laser dye (Marason 1981) it has found use in solar concentrators (Sansregret et al 1983). The structure of DCM is given in figure 1.

It can be considered as a merocyanine type dye consisting of an electron donor (dimethylamine) group linked by a conjugate π electron system to an electron acceptor (dicyanomethylene) group. After excitation with light it has been found to display a strong increase in dipole moment (5.6D to 26.3D) (Meyer and Mialocq 1987). This is due to intramolecular charge transfer, which occurs on much faster timescales than can be observed by the apparatus used here.

Work has shown DCM to exist in both trans and cis isomeric forms (Drake et al 1985), with solutions prepared in darkness existing solely in the trans form. On exposure to ambient light this partially photoconverts forming an equilibrium with the cis form. The ratio of these forms is both solvent and

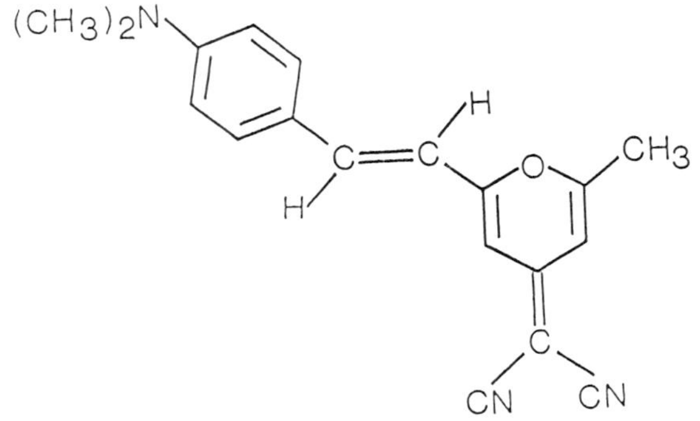

Fig. 1: Structure of DCM

concentration dependent. As the distance between unlike charges is greater in the trans form than in the cis the trans isomer is more stable in polar solvents. It is therefore likely to have a greater abundance and lifetime than the cis (if it fluoresces). The solvent employed has an effect on the spectral properties of DCM, generally the greater the dielectric constant of the solvent the further the red shift in the emission spectrum (Meyer and Mialocq 1987), the absorption spectrum shape being unaffected (Hsing–Kang et al 1985).

Studies have suggested that the fluorescence of DCM consists of two overlapping bands (Hsing–Kang et al 1985), a longer wavelength band appearing as a shoulder on a shorter wavelength band. This shoulder becomes more pronounced in solvents of higher dielectric constant. The intensity of these bands has been found to increase with time and their presence suggests that dual fluorescence found in some solvents comes from two physically different emitting states. A single fluorescence decay is then thought of as originating from these two states in dynamic equilibrium. It has also been put forward (Drake et al 1985 and Hsing–Kang et al 1985) that a geometrically twisted conformation exists between the trans and cis forms, similar to the model suggested for DODCI by Rulliére (1976) shown in figure 2.

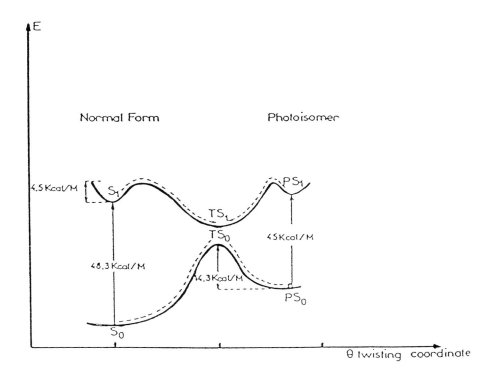

Fig. 2: Diagram showing energy against twisting coordinate for the photoisomerisation of DODCI by Rulliére. S_0-normal form ground state. S_1-normal form excited singlet state. PS_0-photoisomer ground state. PS_1-photoisomer excited singlet state. TS_0-twisted ground state. TS_1-twisted excited state.

Fluorescence is also thought to originate from this twisted intramolecular charge transfer state. Extensive work on these states has been performed by Grabowski and co-workers (1979).

Previous work on DCM in methanol (Meyer et al 1988) has indicated that fluorescence emanates from both the cis and trans forms with just the trans form fluorescing in DMSO. Here, as well as giving more conclusive evidence for cis and trans emission in methanol, we put forward the ideas of either a dual fluorescence for DCM in DMSO or a distribution of decays about a mean lifetime (T_m) as reported by James and Ware (1985) for homotryptophan.

2. Experimental

Experiments were performed using a fluorescence lifetime spectrometer, shown schematically in figure 3. Samples of $\sim 10^{-5}$M DCM in both

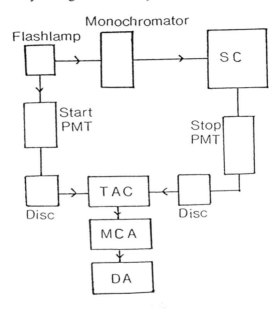

Fig. 3: Schematic diagram of single photon counting fluorescence lifetime spectrometer. TAC-time to amplitude converter, MCA-multichannel analyser, disc-discriminator, PMT-photomultiplier, SC-sample chamber with 570nm cut-off filter on emission, DA-computer for data analysis.

methanol and dimethyl sulphoxide (DMSO) were excited at 480nm with a 20nm bandpass, selected using a monochromator, with light from a coaxial flashlamp operating with hydrogen as a filler gas. Fluorescence over 570nm, chosen using a Schott cut off filter, was detected using a Philips XP2254B (for measurements with methanol) and a Philips XP2257B photomultiplier (for measurements with DMSO). The latter was operated in a cooled housing and its performance has been reported, Birch et al (1988). By the use of NIM timing electronics data was accumulated and displayed on a multichannel analyser (MCA) to the required data precision before being processed using reconvolution analysis on a computer. The data presented in figures 4 and 5

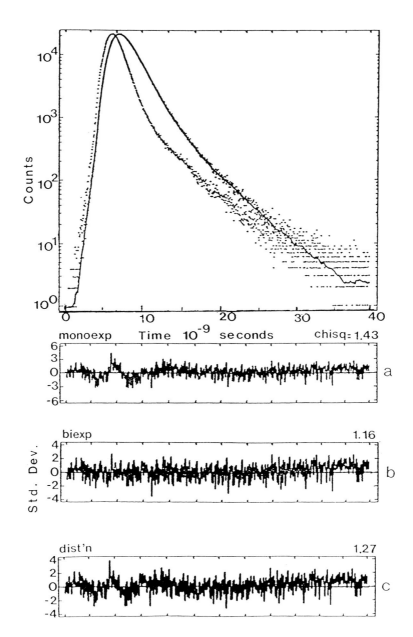

Fig. 4: Fluorescence decay of DCM in methanol, showing decay, fitted function and instrumental response. Residuals are shown for (a) monoexponential analysis $T = 1.38 \pm 0.01$ns, XSQ = 1.43. (b) biexponential analysis $T_1 = 0.32 \pm 0.23$ns, $I_1 = 11.1\%$, $T_2 = 1.44 \pm 0.01$ns, $I_2 = 88.9\%$, XSQ = 1.16. (c) distribution analysis $T_m = 1.33 \pm 0.01$ns, $dT_m/T_m = 0.299$, XSQ = 1.27.

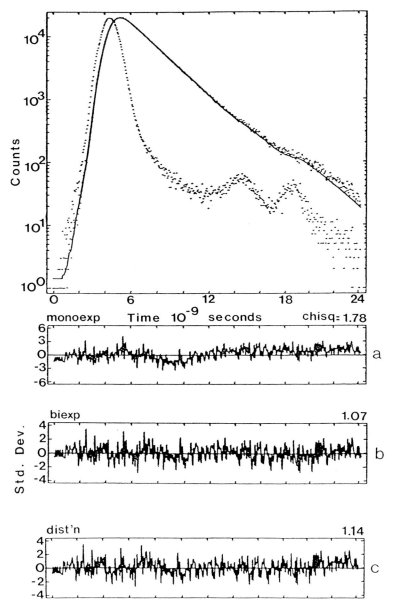

Fig. 5: Fluorescence decay of DCM in DMSO showing decay, fitted function and instrumental response. Residuals are shown for (a) monoexponential analysis $T = 2.15 \pm 0.01$ns, XSQ = 1.78. (b) biexponential analysis $T_1 = 2.04 \pm 0.05$ns, $I_1 = 93.0\%$, $T_2 = 3.62 \pm 0.30$ns, $I_2 = 7.0\%$, XSQ = 1.07. (c) distribution analysis $T_m = 2.09 \pm 0.01$ns, $dT_m/T_m = 0.270$, XSQ = 1.14.

was measured with a resolution of 0.0949ns/channel with 20,000 counts in the peak at 25°C. That presented in tables 1 and 2 was measured with 0.097ns/channel with 20,000 and 10,000 counts in the peak respectively. Both solvents and the cut-off filter were checked for fluorescence before proceeding with the measurements and found not to exhibit any. Fluorescence decays were analysed using both mono-and biexponential decay models along with a distribution of lifetimes about a mean (T_m) model, the goodness of fit of these models being given by the chi-squared (XSQ) value. form having the shorter lifetime and a relative abundance similar to that found by Drake and co-workers (1985). The lifetime data obtained on changing temperature, presented in table 1, gives good evidence that the shorter lifetime does emanate from the cis isomer. One would expect a cis isomer to possess a shorter fluorescence lifetime than the trans form of the molecule. Also, the relative abundance of the cis form would be expected to increase systematically with temperature. Table 1 clearly shows this to be the case, providing new evidence to consolidate the idea of the shorter-lived fluorescence originating from the cis form.

A time-resolved excitation spectrum was carried out by scanning the excitation monochromator, analysing first the rising edge of the fluorescence decay and then the tail of the decay. In the rising edge of the decay one would expect both fluorescing species to be present and in the tail an increased abundance of the longer lived species should be present. A shift to longer wavelength was observed in the spectrum for the tail as compared to that of the rising edge of the decay, thus indicating a strong likelihood that both of the cis and trans isomers exist in the ground state.

b) DCM in DMSO

Here as when using methanol as a solvent we found that a monoexponential decay model did not best describe the decay, although previous work had suggested this (Drake et al 1985 and Meyer et al 1988). In fact in work presented by Meyer and co-workers (1988) a small trend can be seen in the residuals of the data they present for the monoexponential decay analysis for DCM in DMSO. Figure 5 shows the decay of DCM in DMSO analysed using different models. It clearly shows the monoexponential model to be inappropriate. However, there is some ambiguity as to whether the

Table 1. Biexponential decay data for DCM in methanol

Temperature (°C)	T_1 (ns)	I_1 (%)	T_2 (ns)	I_2 (%)	XSQ
−35	0.92 ± 1.21	0.3	2.32 ± 0.02	99.7	1.35
−15	1.02 ± 0.86	3.7	2.08 ± 0.02	96.3	1.14
0	0.62 ± 0.62	6.5	1.82 ± 0.02	93.5	1.26
25	0.38 ± 0.37	7.1	1.39 ± 0.01	92.9	1.01
40	0.31 ± 0.28	9.2	1.16 ± 0.02	90.8	1.34
65	0.25 ± 0.21	19.4	0.85 ± 0.18	80.6	1.18

biexponential or distribution model best describes the decay. This ambiguity in the interpretation of the fluorescence decay curves between discrete and distribution of exponentials was first pointed out by James and co-workers (1985 and 1987). This type of behaviour is now well established in complex systems, e.g. polymers, molecules adsorbed on surfaces and heterogeneous systems (membranes), but has not been widely observed in simple solvent/solute systems.

The effect of changing temperature on the decay parameters is illustrated in table 2. This also emphasises the inability to distinguish between the two decay models used, as the chi-squared values are extremely close. Both models show the expected decrease in lifetime with increasing temperature. Considering the biexponential decay and comparing it with the results obtained using methanol as a solvent, it is not as easy to assign the lifetimes to be originating from both cis and trans isomers. One would expect the cis

Table 2 Data for DCM in DMSO analysed using biexponential and distribution decay models.

Temperature (°C)	Distribution fit		
	Tm ± 0.01 (ns)	dTm/Tm	XSQ
25	2.09	0.229	1.07
35	1.90	0.193	0.98
45	1.74	0.191	1.04
55	1.56	0.225	0.99
65	1.41	0.203	1.02
75	1.25	0.203	1.00

Temperature (°C)	Biexponential fit				
	T_1 (ns)	I_1 (%)	T_2 (ns)	I_2 (%)	XSQ
25	2.02 ± 0.26	90.3	3.16 ± 0.50	9.7	1.06
35	1.86 ± 0.45	84.0	2.45 ± 0.38	16.0	0.97
45	1.69 ± 0.32	88.5	2.36 ± 0.46	11.5	1.02
55	1.43 ± 0.16	65.9	1.89 ± 0.18	34.1	0.97
65	1.29 ± 0.48	59.2	1.62 ± 0.16	40.8	1.02
75	1.20 ± 0.11	86.3	1.69 ± 0.35	13.7	1.00

form, if it fluoresces, to have a shorter lifetime than the trans. However, if this was the case there would be a contradiction between the relative abundances of the fluorescing species presented in table 2 and those obtained by Drake and co-workers (1985). He found that for DCM in DMSO 8% of the molecules exist in the cis and 92% in the trans form. The closeness of the lifetimes recovered show this to be a difficult measurement to perform. Considering the distribution model the fluorescence could originate from a distribution of lifetimes in the trans form caused by a spread of polarisabilities (DCM exhibits a large change in dipole moment) or conformations. It is possible that the fluorescence does originate from two states having similar energies giving the appearance of a distribution.

It should be noted that these measurements, due to the nature of the fluorescence – in terms of closeness of lifetimes and their relative abundance, are not simple to interpret. Further work is under way to distinguish between the distribution and biexponential models in the case of DCM in DMSO and the results will be reported shortly.

References

Antonov V S and Hohla K L 1983 *Appl. Phys.* B32 9
Birch D J S, Hungerford G, Nadolski B, Imhof R E and Dutch A D 1988 *J. Phys. E.: Sci. Instrum.* 21 857
Drake J M, Lesiecki M L, Camaioni D M 1985 *Chem. Phys. Letts.* 113 530
Grabowski Z R, Rotkiewicz K, Siemiarczuk A, Cowley D J and Baumann W 1979 *Nouv J. Chim* 3 443
Hammond P R 1979 *Opt. Commun.* 29 331
Hsing-Kang Z, Ren Lau M, Er-Pin N and Chu G 1985 *J. Photochem.* 29 397
James D R and Ware W R 1985 *Chem. Phys. Letts.* 120 450
James D R, Liu Y-S, Petersen N O, Siemiarczuk A, Wagner B D and Ware W R 1987 *Proc. SPIE* 743 117
Marason E G 1981 *Opt. Commun.* 37 56
Meyer M and Mialocq J C 1987 *J.Phys.Colloque Suppl.* 12 (C7-54)
Meyer M and Mialocq J C 1987 *Opt. Commun.* 64 264
Meyer M, Mialocq J C and Rougée M 1988 *Chem. Phys. Letts.* 150 484
Rulliére C 1976 *Chem. Phys. Letts.* 43 303
Sansregret J, Drake J M, Thomas W R L and Lesiecki M L 1983 *Appl. Opt.* 22 573

Radial model for the optogalvanic effect in the neon positive column

K W McKnight* and R S Stewart, Department of Physics and Applied Physics University of Strathclyde, Glasgow G4 ONG

ABSTRACT

The rate equations describing the optogalvanic effect involve the densities of the interacting particles and generally these vary spatially. This is especially true in positive column plasmas where the electron density has an approximate zero-order Bessel function profile and the metastable and resonance atom profiles may be considerably narrower, resulting in radially dependent laser absorption and excited-atom loss rates. We have included the radial dependencies in the optogalvanic rate-equation model and a final theoretical expression for the optogalvanic current perturbation has been obtained by integration over the total discharge cross-section. The results of our radial model are compared with those of the average model (where the electron and excited-atom densities are averaged over the discharge cross-section) by comparison of the 1s coupling coefficients required to fit the two theories with experimental signals obtained for the 609.6nm sign-change $1s_4 - 2p_4$ transition of neon.

INTRODUCTION

Recently there has been considerable success (Lawler 1980, Doughty and Lawler 1983, van de Weijer 1986, Labat and Bukvic 1988, Sasso *et al* 1988 and Stewart *et al* 1990) in modelling the optogalvanic effect (OGE). This paper considers the effects of the radial variation of the particle densities in the

---

* [Present address: Rex, Thompson & Partners, Newnhams, Farnham, Surrey GU9 7EQ.]

© 1991 IOP Publishing Ltd

318   *Optogalvanic Spectroscopy*

positive column of the normal glow discharge. Previous models of the OGE in the positive column (Doughty and Lawler 1983 and Stewart *et al* 1990) have averaged the particle densities over the discharge cross-section. However, the balance equations used in these models have many terms describing two-body collision processes and these terms involve the product of two radially-dependent particle densities, so radial inhomogeneity is exaggerated. Also, due to the steep radial density gradients, the particle loss rates may vary spatially and these are also multiplied by radially-dependent densities. And of course, if the excited atom densities vary spatially, then so will the laser absorption, the most fundamental quantity in the OGE.

We have considered these effects by modifying the rate equations of Stewart *et al* (1990) to include local values of all the relevant quantities.

DISCUSSION OF THEORETICAL TERMS

(a) <u>The rate equations</u>

To help with the discussion we again write down the electron and 1s atom balance equations (Stewart *et al* 1990):

electron density balance

$$\dot{n} = nS_g N_g + n(S_2 R_2 + S_3 M_3 + S_4 R_4 + S_5 M_5)$$

$$+ \frac{T}{2}(R_2 + M_3 + R_4 + M_5)^2 - n D_a \left[\frac{2.4}{R}\right]^2 = 0 \qquad (1)$$

the $1s_5$ atom balance

$$\dot{M}_5 = nS_{g5}N_g - M_5W_5 - M_5S_5n - M_5T(R_2 + M_3 + R_4 + M_5)$$

$$- M_5 n S'_5(1 - \beta'_5) + n\beta'_5 (R_2S' + M_3S' + R_4S')$$

$$+ n(R_2E_{25} + M_3E_{35} + R_4E_{45}) - nM_5(E_{52} + E_{53} + E_{54})$$

$$+ N_g(R_2K_{25} + M_3K_{35} + R_4K_{45}) = 0 \qquad (2)$$

There are similar equations for the $1s_{2,3,4}$ atoms. ($S'_{2,3,4}$ are the total electron collisional coefficients to the 2p states.) In these equations the symbols are as defined previously.

(b) **The electron density: n(r)**

The radial profile of the electron density follows a zero-order Bessel function (Fig 1)

$$n(r) = n(o)J_o(2.4 \frac{r}{R}) \qquad (3)$$

where $r = o$ and $r = R$ define the tube axis and tube wall respectively. The electrons are lost to the walls by ambipolar diffusion. As before (Doughty and Lawler 1983, Stewart et al 1990) the diffusion term was eliminated from the simultaneous rate equations and so did not require explicit evaluation.

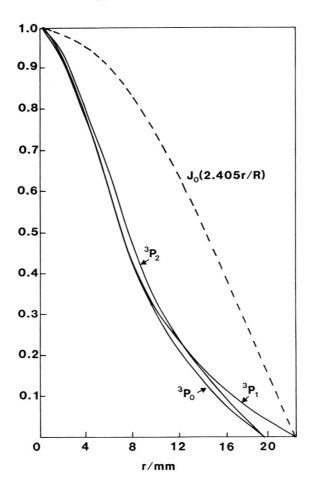

Fig 1 ELECTRON AND 1s ATOM DENSITY PROFILES
(Neon: 6mA, 2 Torr) (Rajotte 1973)

(c) <u>The excited atom densities:</u> $R_{2,4}(r)$, $M_{3,5}(r)$

The 1s atom profiles (Fig 1) are usually somewhat narrower than the zero-order Bessel function profile of the electrons (Rajotte 1973, Steenhuysen and de Brugh 1975, Smits and Prins 1975, Valignat and Leveau 1981). The excited atom profiles are strongly influenced by the electron density profile but are determined in a rather complex way by radial variation of the production and loss processes.

(d) <u>Laser radiation absorption profile: $J_L(r)$</u>

The absorption of the laser photons is probably the most fundamental process involved in the OGE. As seen above, the densities of the 1s absorbing atoms vary steeply with radial position and therefore the laser absorption profile will vary radially. We used annular laser beams, selected from a uniform wide beam, to measure the radial absorption profile $J_L(r)$, shown in Fig. 2. The shape of the curve also depends on the column length. The longer the column the broader will be the profile. Calculations of the laser absorption agreed well with the measured profiles.

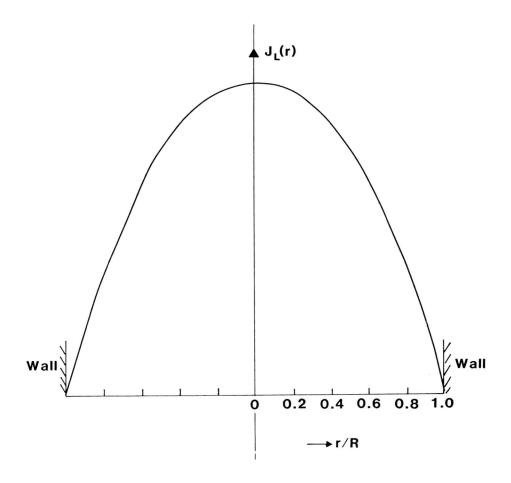

Fig 2   THE LASER ABSORPTION PROFILE

322 Optogalvanic Spectroscopy

(e) <u>Two-body collisions and the density products: n(r).M(r), R(r).M(r) etc.</u>

The radial profile effect will be exaggerated because two quickly varying functions are multiplied. To illustrate this n(r).M(r) and R(r).M(r) are shown in Fig 3. Clearly these products are large for only a narrow region near the column axis.

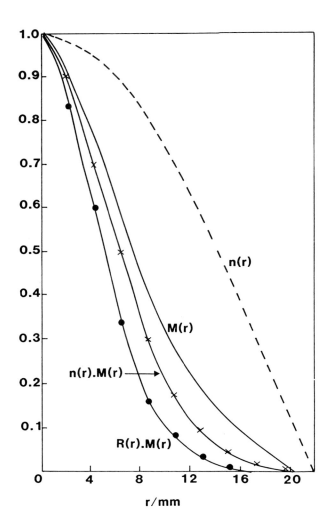

Fig 3   PROFILES OF DENSITY PRODUCTS
(derived from Fig 1)

(f) <u>Metastable atom diffusion frequency: W_M</u>

We tested our model for two different calculations of the metastable diffusion losses to the tube walls. Firstly, we assumed that the loss rate was determined by the fundamental diffusion mode and given by

$$W_M = \frac{D_M}{N_g} \left[\frac{2.4}{R}\right]^2 \qquad (4)$$

where $D_M = (7.6 \times 10^{18} \text{m}^{-1}\text{s}^{-1})(T_g/K)^{0.73}$, (Phelps 1959). Here N_g and T_g are the ground-state atom density and the gas temperature. Previous measurements (Stewart et al 1990) showed that, under our conditions, there was negligible gas heating so the gas temperature was assumed to be constant over the tube cross-section and equal to 300K.

In practice the diffusion loss frequencies depend on the actual radial profiles of the excited atoms which are different from that of the lowest diffusion mode. Therefore we also attempted to estimate the local diffusion frequency $W_M(r)$ by fitting an exponential function to the radial profiles published by Rajotte (1973):

$$M(r) = M(o) e^{-g(r/R)^2} \qquad (5)$$

where the multiplier (g) in the exponent varies with the gas pressure. Published results (Smits and Prins 1975) show that the 1s atom profiles become more constricted with increasing pressure so that g varies from approximately 2 at 1 Torr to about 5 at 8 Torr.

Clearly, equation (5) does not fit the observed profiles near the tube walls, so a fine tuning second exponential was included for large r. In fact, the outer regions of the discharge make a much smaller contribution to the OG signal, so the lack of an exact fit is not crucially important for this estimation. The radial variation of $W_{3,5}(r)$, calculated from

$$\dot{M}(r) = M(r) \cdot W_M(r) = \left[\frac{D_M}{N_g}\right] \nabla^2 M(r), \qquad (6)$$

is shown in Fig 4.

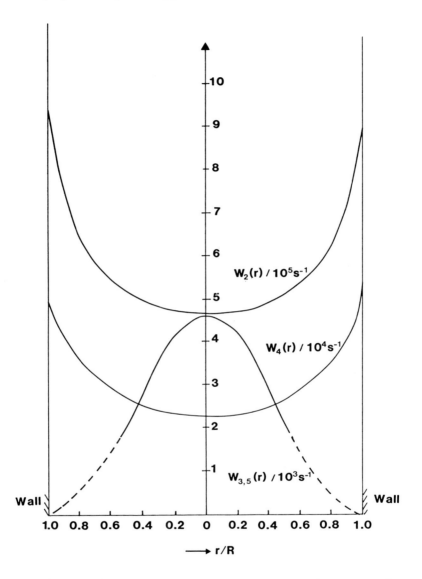

Fig 4   Estimated variations of local loss rates (p = 2 Torr)

(g) <u>Resonance radiation trapping and wall losses of $1s_2$ and $1s_4$ atoms: W_R</u>

Again, the model was tested for two different calculations of the resonance atom loss rates.   First, they were calculated as previously described (Stewart *et al* 1990) by combining the Holstein (1947, 1951) Doppler and pressure broadened loss rates (Vermeersch 1990).

The loss frequencies calculated in this way correspond to the decay of the radial density profile of the lowest decay mode (Holstein 1947, 1951). The actual profiles are considerably narrower (see Fig 1), with the result that the local loss frequency will vary with position. We have attempted to estimate $w_{2,4}(r)$ by using the pressure–broadened loss rate given by Golubovskii *et al* (1971) and the Doppler–broadened rate given by Skurnick and Schacter (1972). Typical radial variations of the calculated loss frequencies of $1s_2$ and $1s_4$ atoms are shown in Fig 4. It should be noted that if there were substantial gas heating this would also affect the radiation trapping loss rates.

We should like to emphasise that the curves shown in Fig 4 are very rough estimations and should be treated with a great deal of caution.

(h) The loss terms: $M.w_M$ and $R.w_R$

Using the first calculations of w_M, these terms will follow the density profiles. However, it is interesting to note that our estimations of the local loss rates $w_M(r)$ and $w_R(r)$ would, in one case, exaggerate the axial behaviour of the product $M(r).w_M(r)$ for the $1s_{3,5}$ atoms but for the resonance atoms, would ensure that the value of $R(r).w_R(r)$ is significant over the whole tube cross–section, because the radial behaviours of $w_R(r)$ and $R(r)$ will effectively cancel.

(i) The optogalvanic kinetic term: Δ

Stewart *et al* (1990) derived a theoretical expression (based on that of Lawler 1980) for the current perturbation Δi and showed that theory and experiment could be usefully compared using the kinetic term Δ, given by

$$\Delta_{exp} = \frac{\Delta i}{\dot{Q}} \left[1 + \frac{Z}{Z_d} \right] \tag{7}$$

where $Z_d = dV/di$, the dynamic impedance of the positive column

and
$$\Delta_{theory} = \frac{\partial i}{\partial n} \frac{K_Q}{K_n} \tag{8}$$

where all the terms are as previously defined.

The quantity K_Q/K_n is a function of the partial derivatives in the perturbed rate equations and all of these equations will contain the strong radial dependence of the terms discussed above, therefore the magnitude of K_Q/K_n will be strongly weighted towards the axial value.

(j) <u>The impedance term: $1 + Z/Z_d$</u>

This term was shown by Lawler (1980) to describe the fact that the positive column is part of the complete circuit. We have considered how this impedance term requires to be mathematically treated when integrating the optogalvanic contributions from annular volumes at different radial positions in the discharge. Due to the low laser chopping frequency (93Hz) which we have employed, we have assumed that the electron density perturbation, caused by laser absorption in any particular annular volume, will be distributed throughout the entire discharge tube according to the fundamental diffusion mode (Golubovskii and Lyaguschenko 1975). Steady-state conditions are established in both the laser-on and laser-off situations, so the electron density radial profile will follow a zero-order Bessel function. Therefore the impedance term will remain the same for all annuli in the integration. We conclude that this term will lead to no radial weighting and therefore may be included along with Δi in the equation for Δ_{exp} (equation (7)).

(k) <u>Calculation of Δ_{theory}</u>

The integrated value of Δ_{theory} is given by

$$\Delta_{theory} = \frac{\int_0^R \left[\frac{\partial i}{\partial n} \frac{K_Q}{K_n}\right]_r \cdot \dot{Q}(r) \cdot 2\pi r L\, dr}{\int_0^R \dot{Q}(r) \cdot 2\pi r L\, dr} \tag{9}$$

In practice we measured the rate of absorption of laser photons by each annulus so we replaced the weighting in equation (9) by the laser absorption profile $J_L(r)$ (see Fig 2):

$$\Delta_{theory} = \int_o^R \left[\frac{\partial i}{\partial n} \cdot \frac{K_Q}{K_n}\right]_r \cdot J_L(r) dr \qquad (10)$$

COMPARISON OF THEORY AND EXPERIMENT

The experimental conditions were as described by Stewart et al (1990). The OGE was studied in the positive column of discharges in neon at pressures from 1.5 to 8 Torr and currents from 1 to 10mA in a cylindrical tube of internal diameter 8.5mm. Typical results are shown in Fig 5 for p = 2.0Torr.

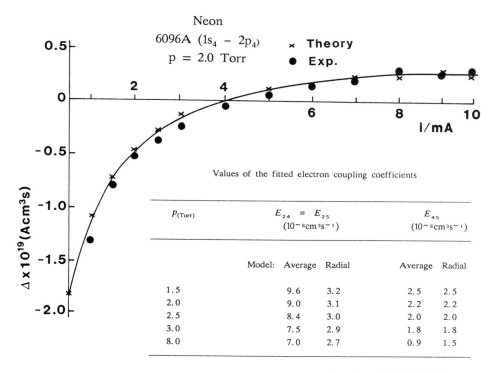

FIG 5   COMPARISON OF THE FITTED 1s ELECTRON COUPLING COEFFICIENTS FOR THE RADIAL AND AVERAGE MODELS

We again used the sign change of the 609.6nm ($1s_4$–$2p_4$) line as a sensitive test of the theoretical model. As previously, the 1s direct electron coupling coefficients were adjusted to fit Δ_{theory} with Δ_{exp}. We did this using the loss rates W_M and w_R calculated for the fundamental mode.

The table in Fig 5 compares the fitted 1s coupling coefficients for this radial model with those of the tube-averaged model (Stewart et al 1990). The $E_{24,5}$ values are considerably lower for the radial model and somewhat closer to the literature values (Steenhuysen 1980).

When we used the local values of the $1s_{3,5}$ and $1s_{2,4}$ atom loss rates as shown typically in Fig 4, this resulted in fitted E values which were about 50% higher. This difference was less than one might expect on first examination. However, this relatively small difference may be due to the fact that most of the important processes take place near the tube axis where the loss rates do not change so steeply (see Fig 4). We conclude that most of the optogalvanic signal is produced in an axial cylinder of radius approximately R/3. Complex calculations of local excited-atom loss rates may not be worthwhile because the model is not highly sensitive to changes in their values.

CONCLUSIONS

We have allowed for the radial variations in the particle densities in the rate equations for our model of the optogalvanic effect in the neon positive column. The model has been tested for the 609.6nm ($1s_4$–$2p_4$) transition for currents from 1 to 10mA and pressures from 1.5 to 8 Torr. The main problem with this radial model is the calculation of the local excited-atom loss rates, although precise calculation of the loss profiles would not appear to be necessary since the outer regions of the column do not contribute greatly to the optogalvanic effect. However, examination of the various aspects of a radial model has given improved insight into the processes giving rise to the optogalvanic effect and indicates that averaging over the discharge tube should be done with care.

ACKNOWLEDGEMENTS

We are indebted to the SERC for postgraduate support and to Mr J Barrie, Mr D Graham and Mr I Wallace for technical assistance.

REFERENCES

Doughty D K and Lawler J E 1983, Phys. Rev. **A28**, 773.

Golubovskii Yu B, Kagan Yu M and Lyaguschenko R I 1971, Opt. Spectrosc. **31**, 10.

Golubovskii Yu B and Lyaguschenko R I, 1975, Opt. Spectrosc. **38**, 628.

Holstein T 1947, Phys Rev **72**, 1212; 1951 Phys. Rev. **83**, 1159.

Lawler J E 1980, Phys. Rev. **A22**, 1025.

Phelps A V 1959, Phys. Rev. **114**, 1011.

Rajotte R 1973, Phys. Letts. **43A**, 47.

Sasso A, Ciocca M and Arimondo E 1988, J. Opt. Soc. Am. **B5**, 1984.

Skurnick E and Schacter H 1972, J. Appl. Phys. **43**, 3393.

Smits R M M and Prins M 1975, Physica **80C**, 571.

Steenhuysen L W G 1980, Beitr. Plasma Phys. **21**, 301.

Steenhuysen L W G and Aan De Brugh H 1975, Physica **79C**, 207.

Stewart R S, McKnight K W and Hamad K I 1990, J. Phys. D: Appl. Phys. **23**, 832.

Valignat S and Leveau J 1981, Physica **104C**, 441.

Van de Weijer P 1986, IEEE Trans. Plasma Sci. **PS-14**, 464.

Vermeersch F 1990 in these Proceedings.

Perturbations induced in CO–He–Xe discharges by laser oscillation

C.E. Little[†] and P.G. Browne

School of Mathematics, Physics, Computing and Electronics, Macquarie University, NSW 2109, Australia

ABSTRACT: An overview is presented of the mechanisms that lead to the large perturbations in discharge current, voltage, gas pressure and sidelight intensities when laser radiation is chopped on and off in the optical cavity of an infrared (5-6 μm) CO-He-Xe (1.3:18.7:2.0) laser. These phenomena give insight into the kinetic processes of CO laser discharges, including information relevant to the dissociation of CO.

1. INTRODUCTION

The CO laser operates at 5-6 μm on a number of rovibrational transitions in the $X^1\Sigma^+$ electronic ground state of CO. For sealed room-temperature CO-He-Xe lasers, efficiencies as high as 5-15% are normally achieved, and hence the stimulated emission is a major sink for energy within the gas discharge. If laser oscillation is interrupted (by obstructing one of the mirrors, for example) then the energy balance of the system is altered and significant changes occur in discharge current (optogalvanic effect) and voltage, gas pressure and the intensities of sidelight of the species CO, Xe, C_2 and CH. These phenomena have been studied to elucidate kinetic processes which occur in the CO laser, which is the prime example of a molecular system where V-V vibrational kinetics dominate over V-T kinetics. We present here an overview of the perturbations induced by oscillation in a CO-He-Xe laser in current, voltage, pressure and sidelight intensities of CO, Xe, C_2 and CH.

2. EXPERIMENT

The perturbations in voltage, current, pressure and sidelight induced by laser oscillation were monitored over a wide range of conditions in a discharge cell which was placed within the optical cavity of a CO laser. The quartz discharge cell was a shortened version of the CO laser tube, having an inside diameter of 10 mm, and was water cooled and fitted with Brewster windows of CaF_2. The length of cell discharge within the cavity was 15 cm. The cell was filled with the same nominal CO-He-Xe (1.3:18.7:2.0) mixture as the laser, but could be operated over a wider range of conditions whilst still permitting lasing in the cavity. The laser radiation in the cavity was chopped by a toothed wheel between the cell and the laser tube. With the high-voltage power supply operated in both constant-current and constant-voltage modes, the cell current and filling pressure, chopping frequency and laser output were in turn varied about the standard experimental conditions while the perturbations were measured. The standard experimental conditions were 15 mA current at 30 torr filling pressure with a chopping frequency of 22 Hz. Perturbations were measured with the laser operating single-line and multiline.

[†] Now at Department of Physics and Astronomy, University of St Andrews, North Haugh, St Andrews, Fife KY16 9SS, UK.

Changes in current were detected as voltage changes across a 67.7 Ω resistor in series with the discharge in the cell. Optovoltaic changes were detected across a 46.6 kΩ resistor which together with a 193 MΩ resistor formed a voltage divider in parallel with the discharge in the cell. Changes in gas pressure in the cell were detected with a differential capacitance micromanometer. Sidelight perturbations were monitored by a 0.45 m Czerny-Turner grating monochromator with a photomultiplier at its output. All perturbations were measured with a lock-in amplifier, with the reference signal derived from the mechanical chopper within the cavity.

3. PERTURBATIONS

All processes identified as playing major roles in the perturbations of current, voltage, pressure and the sidelight intensities of CO and Xe are shown in Figure 1. When the laser cavity is unblocked, stimulated emission acts to depopulate the vibrational laser levels. Reduced CO-CO V-V-T anharmonic pumping from the laser levels lowers the populations of all higher vibrational levels. The rates of discharge heating due to CO-CO V-V-T and CO-He V-T collisions involving CO molecules in high vibrational levels are diminished and the temperature of the gas discharge is reduced in the presence of lasing. Laser action therefore leads to an increase in gas number density in the discharge and in its impedance (Little and Browne 1988).

4. CURRENT AND VOLTAGE PERTURBATIONS

The perturbations in current and voltage are summarized in Table 1. With constant-current excitation, the high-voltage power supply responded to the increase in discharge impedance by raising the tube voltage. In the case of constant-voltage excitation, however, there was a decrease in the current through the discharge cell and its ballast resistor in response to laser oscillation. Consequently, the voltage across the discharge rose (negative V-I characteristic) while the voltage across the ballast resistor decreased. Hence there were both current and voltage changes in response to lasing when constant-voltage excitation was used.

Table 1

Discharge Parameters	Perturbations:	Typical	(Maximum)
Constant-Current Excitation:			
tube voltage		+7%	(+9%)
total gas pressure		+0.03%	(+0.045%)
Constant-Voltage Excitation:			
discharge current		-5%	(-25%)
tube voltage		+8%	(+13%)
total gas pressure		-0.015%	(-0.09%)

Table 1. Current, voltage and pressure perturbations observed in the CO-He-Xe laser intracavity cell under the typical discharge conditions of 15 mA current and 30 torr pressure (Little and Browne 1988). Maximum perturbations observed in the ranges 5-50 torr (15 mA) and 2.5-20 mA (30 torr) are given in brackets. Perturbations are normalized to 10 W multiline laser output and 22 Hz chopping frequency. A positive perturbation indicates an increase when the laser cavity was unblocked.

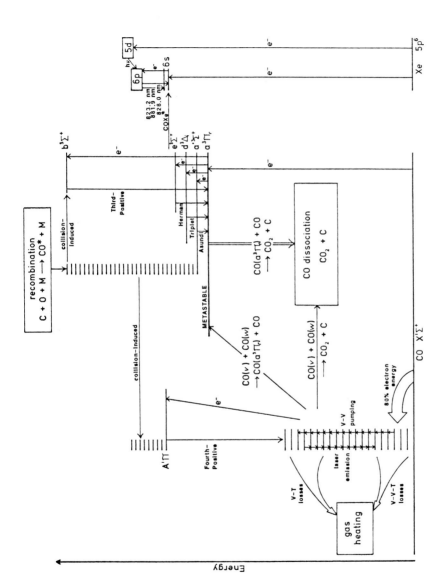

Figure 1. Schematic diagram illustrating the discharge processes pertinent to perturbations induced by stimulated emission in the CO laser.

5. PERTURBATIONS IN GAS PRESSURE

The measured perturbations in gas pressure are also shown in Table 1 above. Although power was removed from the discharge as stimulated emission during lasing, the increase in tube voltage with constant-current excitation caused discharge heating. Thus, the gas pressure in the cell *increased* when laser oscillation began and constant-current excitation was used.

However, during constant-voltage excitation, lasing also caused a decrease in discharge current. The decreases in current and stimulated emission both acted against the increase in tube voltage, and there was a net cooling of the discharge when laser oscillation began. Consequently, gas pressure *decreased* in response to lasing during constant-voltage excitation.

6. SIDELIGHT PERTURBATIONS

Sidelight perturbations observed are listed in Table 2. The perturbations in CO, Xe, C_2 and CH sidelight that accompany changes in laser intensity have all been explained in terms of altered rates of electron impact processes (due to changes in tube voltage, i.e. discharge E/N), and of chemical reactions that are initiated by vibrationally excited CO (whose high-lying vibrational populations are changed by stimulated emission and altered rates of CO-CO V-V-T pumping). Note the large perturbations (up to +190%, -70%) observed.

Table 2

Species	Transitions		Perturbations: Typical (Maximum)	
CO	$b^3\Sigma^+ \to a^3\Pi_r$	Third-Positive group		
		bandhead	-10%	(-25%;+50%)
		bandtails ($N' \sim 30$)	-15%	(-45%)
	$e^3\Sigma^- \to a^3\Pi_r$	Herman bands	not meas.	(increase)
	$d^3\Delta_i \to a^3\Pi_r$	Triplet bands	not meas.	(+130%)
	$a'^3\Sigma^+ \to a^3\Pi_r$	Asundi bands	not meas.	(increase)
	$A^1\Pi \to X^1\Sigma^+$	Fourth-Positive group	+20%	(+60%)
Xe	6p -> 6s	823.2, 828.0, 881.9 nm	+25%	(+190%)
C_2	$d^3\Pi_g(v'=6)$ $\to a^3\Pi_u$	High-Pressure bands	+40%	(+130%)
	$d^3\Pi_g(v'<6)$ $\to a^3\Pi_u$	Swan bands	-30%	(-50%;+90%)
CH	$A^2\Delta \to X^2\Pi_r$	4300 Å bands	-30%	(-70%)

Table 2. Sidelight perturbations observed in the CO-He-Xe laser intracavity cell for the typical discharge conditions of 15 mA current and 30 torr pressure (Little and Browne 1989, 1990). Maximum perturbations observed in the ranges 5-50 torr (15 mA) and 2.5-20 mA (30 torr) are given in brackets. Perturbations are normalized to 10 W multiline laser output and 22 Hz chopping frequency, and were measured with constant-current excitation. A positive perturbation indicates an increase when the laser cavity was unblocked.

6.1 C_2 High-Pressure bands

The intensities of the C_2 High-Pressure bands increase during laser oscillation. These bands arise following three-body recombination of C (Little and Browne 1987), the concentration of which increases during laser oscillation due to an enhancement in CO dissociation:

$$C + C + M \longrightarrow C_2(^5\Pi_g, v) + M$$

$$C_2(^5\Pi_g, v) + M \longrightarrow C_2(^5\Pi_g, v=0) + M$$

$$C_2(^5\Pi_g, v=0) + M \longrightarrow C_2(d^3\Pi_g, v=6) + M \quad (\Rightarrow \text{High-Pressure bands}).$$

6.2 C_2 Swan and CH 4300 Å bands

The intensities of the C_2 Swan and CH 4300 Å bands decrease during laser oscillation, as a result of the suppression of the chemiluminescence reactions responsible for these emissions:

$$C_3 + O \longrightarrow C_2^* + CO \quad (\Rightarrow \text{Swan bands})$$

and $\quad C_2 + OH \longrightarrow CH^* + CO \quad (\Rightarrow 4300 \text{ Å bands}).$

Since the 4300 Å bands depend on C_2 for their production, and the Swan bands arise from the same reaction which produces most of the C_2 in the laser discharge, the Swan and 4300 Å bands are perturbed in the same manner.

6.3 CO Fourth-Positive group

The principal excitation mechanism for the CO Fourth-Positive group has been identified as

$$CO(v) + e^- \longrightarrow CO(A^1\Pi) + e^-,$$

a reaction which is enhanced during laser oscillation. A secondary mechanism

$$C + O + M \longrightarrow CO(a'^3\Sigma^+, v) + M$$

followed by

$$CO(a'^3\Sigma^+, v) + M \longrightarrow CO(A^1\Pi) + M$$

is found to excite preferentially the levels $A^1\Pi(v=1,6,8)$. This process is inhibited during laser action, and leads to reduced perturbations in bands with $v' = 6$ and 8 in particular.

6.4 CO Third-Positive group

Two mechanisms have been identified for the excitation of the $b^3\Sigma^+$ upper state of the CO Third-Positive group. Electron-impact excitation via the $a^3\Pi_r$ metastable state populates primarily the lower rotational levels of $b^3\Sigma^+(v=0,1)$ that lead to Third-Positive bandhead emission. Higher rotational levels that give rise to emission in the tails of those bands are populated via a second mechanism: that of collision-induced crossings from the levels of $a'^3\Sigma^+$ which perturb the higher rotational levels of $b^3\Sigma^+(v=0,1)$. The $a'^3\Sigma^+$ precursor state is populated preferentially in the formation of CO by three-body recombination. During lasing, the electron-impact excitation mechanism is enhanced due to the increase in discharge E/N, while the collision-induced transfer mechanism is inhibited by a reduction in the concentration of O. As a result, for discharge pressures below 15-20 torr, lasing *enhances* the Third-Positive bandhead intensities and *inhibits* those of the bandtails. At higher pressures, the nascent rotational distributions of $b^3\Sigma^+(v=0,1)$ are destroyed by collisions before emission can occur, and both bandhead and tail intensities are inhibited during lasing.

6.5 Xe near-infrared transitions

The upper levels of the strong near-ir Xe transitions in the CO-He-Xe discharge are excited mainly by electron impact from the 6s metastable Xe level, which is itself excited via a COXe* complex from excited CO triplet levels (such as $a'^3\Sigma^+(v=10)$, $d^3\Delta_i(v=5)$ or $e^3\Sigma^-(v=3)$). This excitation route is enhanced by laser oscillation which increases the populations of the CO triplet levels. Excitation of the 6p upper states of the Xe transitions can also occur to a lesser extent by two-step electron excitation via the Xe metastable state, or by electron excitation to higher Xe states followed by radiative decay to the 6p states. These electron-excitation routes will also be enhanced by laser oscillation.

7. DISSOCIATION OF CO IN THE CO LASER

The influence of lasing on CO dissociation is a key factor in many of the sidelight perturbations. By coupling the experimental observations with calculated excitation rates, this work has identified the principal dissociation routes for CO in CO-He-Xe lasers to be the chemical reaction of ground-state CO with metastable $CO(a^3\Pi_r)$ to yield CO_2 and C (Little 1987, Little and Browne 1989):

$$CO + e^- \longrightarrow CO(a^3\Pi_r) + e^-$$

and/or $CO(v) + CO(w) \longrightarrow CO(a^3\Pi_r) + CO$

followed by

$$CO(a^3\Pi_r) + CO \longrightarrow CO_2 + C,$$

and by the direct reaction of two vibrationally excited $CO(X^1\Sigma^+)$ molecules:

$$CO(v) + CO(w) \longrightarrow CO_2 + C \quad.$$

8. CONCLUSIONS

Laser oscillation in a sealed-off room-temperature CO-He-Xe laser has been observed to perturb the discharge current, tube voltage, gas pressure and sidelight intensities of the species CO, Xe, C_2 and CH in the electric discharge. The phenomena have all been explained in terms of the interplay of enhanced rates of electron-impact processes and reduced rates of chemical reactions involving $CO(X^1\Sigma^+)$ molecules in high vibrational levels.

REFERENCES

Little C E 1987 *PhD Thesis* Macquarie University, Sydney
Little C E and Browne P G 1987 *Chem. Phys. Lett.* **134** 560
Little C E and Browne P G 1988 *J. Phys. B: At. Mol. Opt. Phys.* **21** 2675
Little C E and Browne P G 1989 *J. Phys. B: At. Mol. Opt. Phys.* **22** 1269
Little C E and Browne P G 1990 *J. Phys. B: At. Mol. Opt. Phys.* **23** 2433S

Author Index

Allott, R M, *241*
Ashkenasi, D, *289*

Barbeau, C, *65*
Bennett, S C, *191*
Birch, D J S, *307*
Bolovinos, A, *283*
Borthwick, I S, *163*
Browne, P G, *331*
Butler, C O, *191*

Cherry, R I, *297*
Clark, A, *159*

Den Hartog, E A, *27*
Desoppere, E, *133*
Devonshire, R, *207, 257*
Duncan, M, *207, 257*
Dunham, J S, *191*

Freriks, J M, *71*

Goddard, B J, *241*
Greenhow, R C, *81*
Griffiths, T R, *215*

Hakham-Itzhaq, M, *263*
Halewood, J, *81*
Hamad, K I, *89*
Hippler, M, *303*
de Hoog, F J, *71*
Hungerford, G, *307*

Imhof, R E, *307*

Jennings, R, *159, 163*
Jimoyiannis, A, *283*

Jolly, J, *65*
Jones, W J, *215*

Klemz, G., *289*
Kroesen, G M W, *71*
Kronfeldt, H-D, *289*

Lawler, J E, *27*
Ledingham, K W D, *141, 159, 163*
Lins, *233*
Little, C E, *331*

Marshall, A, *159*
McCombes, P T, *163*
McKnight, K W, *89, 317*

Pfab, J, *303*
Pramila, T, *197*

Robinson, T R, *297*

Sasso, A, *169*
Schoon, N, *133*
Shuker, R, *263*
Singhal, R P, *159, 163*
Smith, G, *215*
Snijkers, R J M M, *71*
Stewart, R S, *89, 317*

Telle, H H, *1, 241*
Tsekeris, P, *283*

Vermeersch, F, *109, 133*

Wieme, W, *109, 133*

Yates, A J, *303*